THE LONGSHOREMAN

Richard Shelton headed the Freshwater Fisheries Laboratory at Pitlochry from 1982 to 2001. He is currently Research Director of the Atlantic Salmon Trust and Honorary Senior Lecturer in Environmental and Evolutionary Biology at the University of St Andrews. He lives in Perthshire with his wife Freda.

'I relish those rare occasions when a book I think cannot at all be my sort of thing turns out to be exactly my sort of thing. I enjoyed *The Longshoreman* enormously. Richard Shelton is a fine writer.' Alan Coren

'A magical life on the longshore, alive with natural history.' David Bellamy

'Richard Shelton, not unlike some Victorian naturalists, has successfully combined a delightful autobiography with a scientific history of fishing. He describes an entire world with great charm.' Ronald Blythe

'When I first urged Richard Shelton to write his naturalist's memoir, I never expected him to produce a classic. But he has.' Redmond O'Hanlon

'*The Longshoreman* is a treasure. It is one of those rare books that transcends classification: a sport of nature, a singular success. It is in part boyhood memoir, told with the astonishingly clear recall of small children (the author was terrified of foxes). It is an informative book of natural history, written with the easy charm of the great Victorian classics. It is also the story of a long, happy and successful career in environmental research and the fishing industry. But above all, it is a song of praise to the wonders of fish. Richard Shelton writes of fish with the pen of a poet. Ted Hughes himself could not equal some of these descriptions. The beauties and oddities of the shoreline and the marine world are brought before our eyes in vivid colour and with scientific precision...The whole of this short and delightful book is full of the most fascinating gobbets of information... Oh lucky, gifted man.' Margaret Drabble, *Country Life*

'Minutely detailed and utterly fascinating excursions into natural history and shooting... communicated with an artist's skill. Every adventure and scrape of his professional and personal life, every key incident and discovery, is given immediacy and involvement because Shelton describes it lightly, in lay terms, in the present tense... *The Longshoreman* is a delight. It is also an education. It shows vividly how one man can do the small things well – and how Men can do the big things badly.' Brian Clarke, *The Times*

'Why did I love it? You learn stuff without realizing it, like when you get seated next to a particularly good-value dinner guest. Shelton turns a phrase satisfyingly well; his memory for intricacies is remarkable, remembering buckled sandals and shapes of nets; having cooled Carnation milk in too-strong tea aboard a trawler, the "white china bowl" in which he saw his first eel. How he remembers such details, I do not know, but I am glad he does.' Annalisa Barbieri, *Independent*

'A sensory pleasure to rival the writing of Elizabeth David on food… the great joy of this book lies in Shelton's anecdotes of his time off from the weightier concerns of his job; of a life of brief, luminous moments found in streams, marshes and leaky wellies. These episodes, like a rich aspic, settle over Shelton's life story and leave its rare savour with the reader long after the book has been put down.' Peter Nichols, *Guardian*, Book of the Week

'A most timely and modern reminder of what we stand to lose in our everyday relationship with the natural world... Quirky, passionate, free-ranging and outspoken... it engages your gear.' David Profumo, *Daily Mail*

'Such a fine writer… The result is a book that is not only a well crafted autobiography but which does for marine biology what David Bellamy did for botany, making what might seem a forbidding subject come alive. I simply would not have believed that anyone could write so entertainingly about the toenails of the lobster, or the nasal hairs of the brown shrimp... A pure delight of a book.' Charles Duncan, *Scotsman*

'In a different league to most fusty memoirs of working life. The tides surge through the pages. If you have any appetite, either literal or intellectual, for the mysterious creatures that surround our coast, there is much of interest here... Shelton has produced an uplifting book about a happy life.' Christopher Hirst, *Independent*

'A charming, poetic, funny, vividly descriptive, fascinating anecdotal account, written in pellucid prose by a man of impeccable taste and photographic recall.' John McEwen, *Oldie*

'A beguiling autobiography.' Russell Davies, *Sunday Telegraph*

'Richard Shelton's life is testament to the power of an idea which has, on the whole, been abandoned: immersion in the natural world is the greatest gift that could be given to any child... One of the deep pleasures of the book is the coexistence of the young boy's pure sensory entrancement with the rational understanding of the mature scientist... The method is anecdotal, discursive, vivid, occasionally vulgar and loosely chronological. Never made quite explicit, but hinted at now and then, is the suggestion that water, fish and rivers, and then the sea marshes and their beautiful, shootable geese, and perhaps the sea itself, were a form of refuge for the young bespectacled Shelton... It is almost as if the book were written by a man who had run away to sea and stayed there.' Adam Nicolson, *Times Literary Supplement*

'*The Longshoreman* charms from the first effortlessly beautiful sentence recalling his boyhood in the 1940s but resonates, modestly but authoritatively with a soft spoken eloquence to the final page. He writes... with tenderness, curiosity and sensitivity that recall the best of the classic naturalist writers.' Iain Finlayson, *Saga magazine*

'A classic natural-history memoir.' Giles Foden, *Condé Nast Traveller*

'This is the fascinating memoir of a dedicated but unusual public servant whose tales of life aboard ship are worthy of the saltiest sea-dog.' Richard Knight, *Time Out*

'This eloquent and moving book traces his lifelong love affair with fish, and presents a profoundly sensitive picture of our changing relationship with the seas... At a time when most people's encounters with fish are from the other side of an aquarium window or are mediated by the supermarkets, Shelton takes us into the fishes' world – a world that our voracious appetites are fast driving to the brink of extinction.' Sarah McCarthy, *Ecologist*

'A beautifully written book from a fisherman, wildfowler and eminent marine biologist.' *Shooting Gazette*

'An engrossing mixture of anecdote and theorizing, briny-flavoured throughout and narrated in a style that fishtails between rhapsodic and tight-lipped... Images of fish, family, boats and dead geese embedded in the text (as in the books of W. G. Sebald) and they add to the charm and singularity of a biography that might have been written in saltwater.' *Yorkshire Post*

THE LONGSHOREMAN

A Life at the Water's Edge

Richard Shelton

Atlantic Books
London

First published in Great Britain in hardback in 2004 by
Atlantic Books, an imprint of Grove Atlantic Ltd

This paperback edition published by Atlantic Books in 2005

1 2 3 4 5 6 7 8 9

A CIP catalogue record for this book is available from the British Library.

ISBN   1 84354 162 9

Printed in Great Britain by Mackays of Chatham

Atlantic Books
An imprint of Grove Atlantic Ltd
Ormond House
26–27 Boswell Street
London WC1N 3JZ

For my family and shipmates, past and present

CONTENTS

A Word of Explanation xiii

The Longshoreman xv

Select Reading List 331

Picture Credits 335

A WORD OF EXPLANATION

It is given to few to spend a working life pursuing a childhood passion to its limits. That I was able to do so as a fishery scientist and wildfowler has been the greatest of privileges. When my friend, the travel writer Redmond O'Hanlon, suggested that I compile a sort of memoir of my life at the water's edge, I readily agreed. It would be a way of giving thanks for my good fortune and acknowledging the great debt I owe to the many fishermen and longshoremen I met along the way.

The result is not a complete account, still less is it a book about fishery science. Rather, it is the story of how a small boy's interest in the natural world, steam locomotives and old guns was given free rein by a benign Providence. So far as possible, I have set down the selected personal experiences which form the bulk of the book in the order in which they happened. To these passages I have added what I hope is just enough historical and explanatory material to make sense of a more than usually varied career.

Where I refer to particular freshwater and marine fish and shellfish, I do so using their commonly accepted names with occasional nods in the direction of the vivid, if less polite, ones sometimes given to them by fishermen. To avoid confusion, I have also tried in every case to supply the Latin names of individual species. Unlike vernacular names, these scientific ones normally appear as two words. The first name, which always begins with a capital letter, defines the 'genus' to which a certain organism belongs. The second, or 'specific' name, which is not dignified by a capital, assigns it to an individual 'species'. It is a universal convention that both Latin names are printed in italics, but that the name of the naturalist who originally described the creature appears afterwards in Roman script.

Over the years, the names of many organisms have been changed by successive taxonomists, so the inclusion of the authority who originally gave a particular fish or shellfish its specific name also has to be shown to avoid ambiguity. In a further twist, which is the despair of editors and typesetters the world over, the authority's name is shown without brackets if both the generic and specific names are those of his original description. If, as commonly happens, the beast has since been ascribed to a different genus, the authority's name is shown within brackets. With the single exception of Carl Linné or Linnaeus, the eighteenth-century Swedish naturalist who invented the system of binominal nomenclature, the names of authorities are shown in full. Linnaeus named so many organisms that his name is normally abbreviated to 'L.' and I have followed this convention.

One of the lessons a new author learns when writing a book is just how much of a team effort it all is. I have never had the slightest interest in typewriters or computers and, had I not had such a charmingly patient wife to interpret my scrawl, the book would never even have been started. I have also been fortunate in the editorial skills of my publishers, especially those of Angus MacKinnon, who even found time to make a splendid pencil sketch of the light cruiser HMS *Birmingham*. To him, to his colleague Clara Farmer and the many others whose countless personal kindnesses made the book possible, I say a very big thank you.

RS, September 2003

THE LONGSHOREMAN

'Only a miller's thumb' – my brother Peter fishing in the River Chess

FISHY BUSINESS

The Chess is a chalk stream, one of several in south-east England that help to sweeten the waters of the lower Thames. It rises in the Chilterns in about half a dozen little brooks and winter bournes, fed by springs that bubble out of the ground like liquid crystal. The largest of these brooks flowed close to a sawmill, no doubt long since closed, below which the infant Chess opened out into a long pool shallow enough for small boys to fish in without drowning.

My brother Peter and I are standing by the pool as our mother maintains a watchful eye. Below the surface hang the grey-brown forms of the 'banny-stickles' or three-spined sticklebacks, *Gasterosteus aculeatus* L., jerking forward when they see us and then hanging again, maintaining position with their quivering pectoral fins. Here and there we see the gaudy magnificence of a 'cock fiery' or mature male three-spined stickleback. He is turquoise above and brightest scarlet below, and he is fanning his nest in which several females have been chivvied into depositing their eggs.

Peter and I are carrying home-made nets and we make many clumsy attempts to catch the tiddlers from the bank. How easy it looks, the fish almost stationary before the plunging net. Back the net comes and we search its folds, finding nothing but a little sand and a couple of toe-biters, the freshwater amphipods often, wrongly, called freshwater shrimps. Eventually, my brother is rewarded by the flapping silver of a tiny fish and pops it proudly into the jam jar to which my mother has tied a carrying handle of string. I do not have an immediate tantrum but instead step into the water in the hope of achieving success of my own by confronting the quarry in its element.

*My elder son, John, fishing in
the Kinnesburn, St Andrews*

There's a sudden chill as water which has not long sprung from its cool fastness in the chalk enters my left wellington. I say nothing for fear of bringing the expedition to a premature end and stand stock-still as a trickle into my right boot gathers strength. Slightly raising my eyes to look towards the deeper water in front, I see a seemingly transparent grey ghost gliding into view. It pauses briefly but, before I have time to get over my wonder, it spots me and, magically, it is no longer there.

For some reason, known only to the mercurial mind of a little boy, I tell no one. In fact, I have seen my first trout, *Salmo trutta* L., and in due course I will learn the secret of its transparency. Like those of most mid-water and surface-living fishes, the scales of trout are faced with silvery crystals of guanine, and it was the bio-physicist Eric Denton who first demonstrated that the crystals are arranged in rows which, when parallel with the sun's rays, act as mirrors. By reflecting their surroundings, they give an illusion of transparency and thereby hide the fish.

It proves impossible, however, to hide the fact that I have filled both boots with the icy water. I am still fishless and, to avoid a scene from her spoilt elder son, my mother, revealing a skill which aston-ishes me, deftly catches some tiddlers and pops them into my jar. Boots are emptied and the jars, along with their precious contents, are taken home and put on the sill of the kitchen window. The little fish are admired until it is time for bed. Sadly, though, there is only so much oxygen in the jars and by morning all the fish are dead. But they have already exerted a powerful fascination on both Peter and me. More must be found, and quickly.

ONE SUNDAY MORNING

It is a highland Sabbath and a hesitant sun dapples the gravel around the porch of the tiny kirk. A few cars have already arrived and, along the road, little groups sharpen their pace as time for morning service approaches. For many, youth is a distant memory but children's voices can still be heard under the oaks. Here in the Perthshire hills, the pop culture of the towns has yet to dull the minds of the young folk and the embers of older values still glow brightly in the hearts of the young mothers from 'up the glen'. The minister arrives, a slim, white-haired figure, elegant in black, dark eyes shining out of a face that still has the power to enchant. A reassuring glance here, a kindly word there – she is among her people and any that 'swithered' about turning out today are thankful that their better selves have prevailed.

The wee kirk started life as a mission hall. The simplicity remains – rows of pews, a pulpit, a lectern and a communion table raised up on a low stage, are all that distinguish it from a garage or a 'tatty' (potato) shed. It's different today, though, because over the communion table is draped a full-size blue ensign, in the top left the brilliant complexity of the Union flag, and in the centre of the blue 'fly', the bright oak leaves and crown of the Scottish Fishery Protection and Research Flotilla.

It is Sea Sunday, a time when congregations across Great Britain and Northern Ireland remember that, however far from the sea we may live, we are an island people. The Articles of War, first articulated in the seventeenth century, begin with the words: 'It is upon the Navy, under the good providence of God, that the welfare and safety of this kingdom do chiefly depend'. In her Call to Prayer, the Reverend enchantress commends all seafarers to the care and protection of Almighty God.

Long extempore prayers are part of the Presbyterian tradition. At their longest and most discursive, they test the concentration, if

The blue ensign flown by
Scottish Fishery Protection and Research Vessels

not of the Deity, then certainly that of the most earnest of His worshippers. But this morning no one grudges the seamen of the Royal Navy, the Merchant Service, the Fishing Fleet and the Royal National Lifeboat Institution their separate mentions. The first hymn, 'Will your anchor hold', a favourite of the fishing communities of the Costa Granite (the Moray coast in north-east Scotland) and the 'regimental march' of the Boys' Brigade, thunders out, an involuntary descant supplied by the loud, sharp and quavering voice of an elderly lady to my right, ample of bosom but stout of corset. As we sit down, a voice whispers in my ear, 'Are you nervous?' With an uncertain shake of the head, I make my way to the lectern. As the only working seafarer in the parish, I have been asked to read the Old and New Testament lessons. Despite my rare appearance in the pew, the enchantress has indulged my request to read both of them in the incomparable English of the King James Version. Making my way up to the lectern, I look for reassurance at the blue ensign now draped on the communion table and 'won' many years before with the connivance of a sympathetic Marine Superintendent who looked the other way at just the right time.

'They that go down to the sea in ships, that do business in great waters. These see the works of the Lord, and his wonders in the deep': the familiar words of Psalm 107 bounce back off wall and window-pane alike, and in my nervousness the reflected tone sounds to me like that of a stern and aged headmaster. Back to the pew, more hymns and prayers, a sufficient break to recover in time to struggle through the Gospel account of the stilling of the storm. A thoughtful sermon from the enchantress follows, delivered in the soft cadences of her Aberdeenshire calf country (old Scots for where she was brought up). The final hymn, 'Eternal Father, strong to save', favourite of ships' companies throughout the English-speaking world, nearly breaks me, so often have I heard it sung by my fishermen shipmates, but I reach the end without recourse to my handkerchief.

How had I, born far to the south in Aylesbury, the county town of Buckinghamshire and as far from the sea as any town in England, found myself representing the seafaring community in a village kirk in rural Perthshire? It is a long story, and it is not only about the sea.

<hr>

LUCKY BONES

<hr>

I was born in the mid-summer of 1942. Tobruk had fallen and Alamein was still in the future. My father had joined the Local Defence Volunteers (later to become the Home Guard) immediately they were formed but, as an older married man, had yet to be called up. He was shortly to join the Royal Air Force. As a professional photographer in civilian life, he served as a photographic specialist with Coastal Command. For a time he was based at RAF North Coates in Lincolnshire and, with my mother and me, was billeted at a farm near the aerodrome. First memories are as much a product of brain development as of external events. Thus, I have no recollection of the enormous explosion which accompanied the collision of two fully bombed-up Lancasters near the farm. It must have been some bang because the blast brought the ceiling down in the bedroom containing my cot, from which my mother had removed me moments before. I do, however, remember a yellow tanker lorry which used to visit the farm and a steamroller which worked on the local roads and no doubt also on runway repairs.

Shortly afterwards, my father was posted to the Middle East and my mother and I went to stay with her parents in Aylesbury. Here my early interest in steam propulsion was reinforced by the fact that my grandparents' house overlooked the Aylesbury to Cheddington line, the world's oldest branch line and whose documents of Royal Assent carry the cipher of HM King William IV. Light passenger traffic and some goods were handled during the war by archaic and feeble-looking four-coupled tank engines with very tall chimneys. I dare say they were more capable than they looked to a little boy. The apple of my eye was a magnificent eight-coupled goods locomotive, built originally for the London and North Western Railway, and often to be seen sizzling quietly from my grandparents' newly decorated front room. 'One two THREE, four, one two THREE, four' barked fiercely from its chimney as it took hold of its squealing trucks in the curiously named Dropshort siding opposite the house. I had been given some wax crayons by an honorary aunt and it was not long before a large representation of the engine adorned one cream-distempered wall, a result I had achieved by standing on the sofa. My maternal grandmother was a saint and I was never aware of her disapproval.

Not long afterwards, my uncle, who was serving with the Royal Army Medical Corps, caught meningitis. He had joined the Territorial Army before the war and was called up at once. He was attached to a Territorial battalion of the Royal Fusiliers, a collection of East End reprobates whose fearsome reputation for fighting with one another and with the men of other units had earned them the informal title of the Hackney Gurkhas. Thanks to a new drug developed by May & Baker, my uncle's life was saved but he was declared no longer fit for overseas service. As a result, he was able to come home often enough to play a big part in my early upbringing. My interest, and that of my cousin Stuart, in steam locomotives found a new expression in the lovely wooden toy engines he built for us. Further reinforcement came from a superb double-page illustration of express engines in a volume of the *Children's Encyclopaedia* which had originally been bought for my mother and her younger sister. The engines were resplendent in the gleaming liveries of the independent railway companies that pre-dated the amalgamations following the First World War. My particular favourite was a North Eastern Railway Pacific in the light green of that fine old company.

Thinking back, I often wonder if my interest in natural history had its first flowering in this early obsession, one I have never entirely lost, with a form of propulsion which creates such a convincing expression of a living and breathing organism. As it was, my real biological observations were concentrated on the enormous bumblebees and brightly coloured butterflies that often visited my grandparents' back garden to enjoy the golden-yellow flowers of the monkey musk and the tall blue lupins.

VE-Day came and my cousin and I were given tall paper hats with Union flag motifs together with small Union Jacks with which, despite the outbreak of peace elsewhere in Europe, we duelled whenever our respective pushchairs drew in range of one another. My father's service with the RAF extended into 1946, by which time his travels had taken him to Egypt, the Sudan, Lebanon and Kenya. One day when he arrived home, enormous in his blue greatcoat, I was playing with a small fishing rod, attempting to hook a metal fish out of an enamel bowl. 'I'm fishing' were my words of greeting and his face beamed.

As my younger brother Peter and I gained in understanding, so we pestered our father more and more for stories of his time abroad. He had brought many souvenirs back from Africa: soapstone carvings of fish and elephants from the Nile, an ivory paper-knife in the form of a crocodile, vultures of horn and Kenyan tribesmen and ducks made of wood. Pride of place in the collection was held by the 'lucky bones' of a lion and a leopard. I have them in front of me as I write and they appear to be collar-bones which, in the cat family, float free in the muscle tissue of the upper shoulder. Their reputation as lucky is probably a throw-back to the days when warrior tribesmen proved their manhood by killing lions while armed only with spears and were regarded by enlightened white hunters as 'the bravest of the brave'.

With the souvenirs at hand and my father's skills as a raconteur, we soon became familiar with a world in which hyenas whooped outside the tents and lions roared menacingly in the night. Curiously, none of these stories caused us to lose sleep – none, that is, bar one: the story of how once, one night in the Egyptian desert, he had seen a mountain fox peeping above a sand dune. Exactly why it was that the thought of seeing a small and timid animal with large ears should have engendered such terror was not clear then and still is not nearly sixty years later. The dark was a time when foxes were abroad and we even feared their imaginary indoor presence in my father's photographic darkroom. What nonsense it was, but how vividly the memory remains.

*My mother in 1937 at the wheel of
my father's Austin Seven Ruby Saloon*

During his service overseas, my father's car, a 1937 Austin Seven Ruby Saloon with running boards but no bumpers (they were an optional extra for which my careful father was unprepared to pay), rested on blocks in the coach-house of his mother's property in what was then the small village of Stoke Mandeville. It lay some twelve miles away from our town flat in Chesham. To reach it involved a journey through the Chiltern Hills and on to the southern edge of the Vale of Aylesbury, a wild paradise visited by golden plover in the winter and made famous as Londoner's Leicestershire by the Whaddon Chase hunt. Petrol was still in short supply and some of my first journeys to my paternal grandmother's house were by bicycle. My father had two bicycles, giants with twenty-eight-inch wheels and large frames in proportion. To the crossbar of one he attached a wooden seat which he had made himself. Here I sat as we bowled along, the three-speed gears and the strength of my father's legs – honed by many a run along the perimeter track to service the cameras of Bristol Beaufighters and De Havilland Mosquitoes at dispersal – overcoming all gradients. One day in early autumn, we passed a stack-yard with a threshing machine in action, power provided by a traction engine which was not in the elaborate ex-works livery of the modern traction engine rally but workmanlike in faded black as all of them were at that time.

Before the war, the garden at my grandmother's house had been something of a showpiece. Herbaceous borders surrounded the house, and there were a large orchard and kitchen garden, lawns and a tennis court. By the time my brother Peter and I knew it, nature had had rather a free rein. The herbaceous borders were still in fair order and the kitchen garden had not completely lost the battle against field horsetail, ground elder and rabbits; the lawns were still kept up, but the tennis court had become a meadow. The apples had not been pruned for over twenty years and had become full-size forest trees below which cow-parsley and wild horseradish grew strongly. A better place for small boys to introduce themselves to the natural world could not have been imagined.

The author on the footplate of a steamroller,
North Coates, Lincolnshire, 1944

On a bright summer day in the late 1940s, my brother and I are on our knees on what my grandmother still calls the tennis court. Vetches embrace the soft grass of early summer with their fine tendrils. There is a gentle hum from her bees visiting the red clover and here and there common blue butterflies probe for nectar, their undersides wondrously speckled like a trout with wings. Across in the kitchen garden the rusty outline of a cat, set on a wrought-iron peg, creaks as a slight breeze gets up. It has watched over my grandmother's vegetables since 'Dig for Victory' days. The woodpigeons, which have built their fragile nest in the top of the tall hawthorn near the cesspit, have long grown used to it, and it offers no defence against the rabbits and Cabbage Whites. The latter are so common that my brother and I ignore them. Then one of us spots a Large White, and it looks like a male. We rush over and watch it settle on one of the young cabbages, then stalk cautiously to where we last saw it. At last we see the tips of its wings above a leaf. Just as I am ready to pounce, it takes wing and the gentle breeze lifts it high over the apple trees and away.

Quite unreasonably, I blame my brother for this unsatisfactory outcome, but then my grandmother calls us in for our mid-morning cocoa. She has just lit the Primus stove. Two days' worth of potato peelings are lifted into a blackened old saucepan and soon the scent of their cooking mingles with that of the burnt methylated spirit used to get the stove going. We finish our cocoa and watch the peelings steaming in the chipped old sink. My grandmother chops them up and adds bread crusts and a small amount of a commercial poultry meal. When all is cool, we go with her to feed the hens, two Rhode Island Reds called Alice and Edith. On the way, we quietly sample the food ourselves; it is by no means bad, the flavour of cooked potato and the granary scent of the meal making a richly nutty combination.

The hens live in a wooden hen-house set in a large run under the trees. We tip the mixture into their trough and, while the hens peck avidly and my grandmother tops up their water bowl, my brother and I search the hen-house for eggs. We find two and take them proudly back to the house. Both have thick brown shells well able to resist cracking when they are boiled over the Primus the following morning, and we enjoy the rich golden yolks and incomparable flavour that only naturally produced eggs possess.

My grandmother looks up at the grandfather clock in the living room: 'Quick, boys, it's time for *The Master Cutler*.' Stoke Mandeville station is a field and the entrance of Mr Tapping's neighbouring farm away from the house. Hurriedly we exchange our wellingtons for buckled leather sandals and run all the way to the station. Here and there the neatly slanted fence is punctuated by cast-iron notices declaring the suzerainty of the Metropolitan and Great Central Joint Committee and threatening trespassers on the line with a fine 'not exceeding forty shillings'. We run into the booking hall. There is no nonsense about platform tickets. The friendly stationmaster, Mr Spittles, who had lost a son when HMS *Hood* was sunk by the *Bismarck* in the Denmark Strait, welcomes us warmly but tells us not to stand too close to the edge of the platform.

The lower quadrant signal falls with a clunk and two pairs of eyes peer under the road bridge and down the double-track main line from Aylesbury. Suddenly, a black speck appears where there was none before. The gleaming rails start to sing and, in no time at all, the smoke box of the engine of the Sheffield to London express is thundering under the bridge. The engine, bearing the proud headboard of *The Master Cutler*, is in the oil-streaked apple green of the London and North Eastern Railway and has not long been turned out from their Darlington or Doncaster works. It is a B1 class mixed-traffic locomotive designed by the dour but able Edward Thompson. The varnished teak carriages, commissioned by his predecessor Sir Nigel Gresley, sweep past. Our heads swivel until all we can see is the tail-light of the guard's van as the train climbs to Wendover. For the first time we hear the separate beats of the exhaust as the driver lengthens the cut-off to lift the train up the gradient. Oh, to be an engine driver!

A rod alongside the track moves and the signal arm returns to the horizontal. *The Master Cutler* is making good time and we do not have long to wait before the signal arm falls again. We look towards Aylesbury. A speck reappears, moving more slowly this time, and the black of a smoke box is exchanged for the apple-green bunker of the modern tank engine hauling my aunt's train from Aylesbury. The train draws in, the carriages veterans of the old Metropolitan Railway and known, from the date of their design, as 'dreadnoughts'. Completely ignoring our aunt, we rush up the platform to look at the engine.

We would have preferred a tender engine, but a tank is better than nothing and we can see that this is a new one, despite its want of a good clean. We ram our fingers into our ears as the safety-valve lifts deafeningly but still hear the clank of the fireman's shovel. Grey smoke appears from the chimney, the safety-valve closes, the guard's whistle shrieks, the engine answers and the driver opens the regulator. A black from the freshly coaled fire lands in the corner of my eye. 'What a fuss about nothing,' says my aunt as she deftly removes it with the edge of a spotless white handkerchief and escorts us back for lunch.

In one corner of my grandmother's garden was a wooden-framed cart to which was attached an early stationary gas engine. It had belonged to the engineer uncle I never met; I dare say he had used it to power a dynamo. My brother and I knew it just as the 'old engine' and, in our young imaginings, it had all the glamour of the locomotive at the head of *The Master Cutler*. After lunch we lugged the contraption on to the overgrown tennis court and behind it assembled a 'train' of old packing cases. We attached a wide drainpipe chimney to the front of the old engine, lit a fire at the bottom (we had found some matches in the kitchen) and spent the afternoon rapturously between Sheffield and Marylebone.

That night we went to bed content but, despite a bath, smelling strongly of smoke. Even here, we could enjoy the sounds of the night traffic on the line. The express goods trains were hauled by Green Arrow heavy mixed-traffic locomotives designed just before the war by Gresley. They had three cylinders served by his conjugated valve gear, which worked well enough when it was new or freshly overhauled, but produced a curiously syncopated result when worn. The poor Green Arrows passing my grandmother's had made a splendid contribution to the war effort, hauling troop trains and munitions and playing a big part in the preparations for D-Day. But many had not had a decent overhaul since. Instead of six even beats, 'one-TWO, three, one-TWO, three' enlivened our rest, as did the grandfather clock outside the bedroom door which, after a preliminary whirr, struck the hours with a tremendous '*boing*' that still rings in my head.

Not all of the night sounds were so benign. There was a row of elms along a hedge in the next field and it was a favourite of tawny owls. Their call, 'Whoo-hoo, whoo-hoo', sometimes followed by unfriendly screechings, struck immediate terror and once I fancied I heard a vixen scream. Could there be a greater horror? We were already fearful of foxes. Unwittingly, my grandmother reinforced our anxieties because she often used to sing to us before we went to bed and her favourite song was 'A fox jumped up on a moon-light night'. Nothing is more fascinating than something of which one is afraid, especially when it jumps up. But there was worse. Above the chimney-piece in the living room, my grandmother kept a nearly complete set of the *Badminton Library of Sports and Pastimes*. How often I used to glance at the etching on page 63 of the volume on hunting which, over the legend 'In search of supper', depicted a fox viewed from behind and about to pounce on what looked horribly like Alice and Edith.

The Badminton Library of Sports and Pastimes,
inspiration for many a boyhood exploit

Mr Loveday and his footplate companions,
Peter (left) and the author (right)

My father, who had quietly taken note of our interest in steam locomotives, now told us about his friend, Mr Loveday. A fireman on the Chesham shuttle, Mr Loveday had agreed to let two small, engine-obsessed boys join him for a run on the footplate. The little Chiltern town of Chesham was served by a branch line of the old Metropolitan Railway which joins the main line at Chalfont, and the shuttle consisted of three ancient coaches (they are now preserved as the Chesham Set on the Bluebell Line) originally built for electric haulage. In Mr Loveday's time, power was provided by elegant Atlantic tank locomotives designed by J. G. Robinson and built for the Great Central Railway at their Gorton works in the early 1900s. How we contained our excitement until that greatest of all days I shall never know.

We had badgered my father into arriving very early. The Ruby Saloon drew into the station car park; the Chesham Set was alongside the platform but there was no sign of Mr Loveday or of the tank engine. Surely he had not forgotten? Then we saw the little engine simmering in the goods yard, its tall steam dome even higher than the roof of the cab, and watched as it puffed round the train to take on water. Mr Loveday clambered up and directed the slack leather tube of the water crane into the right-hand tank. In no time it was overflowing, the water crane was swung aside and Mr Loveday was beside us, greeting my father and being introduced to his two wide-eyed footplate companions.

There was one last duty to be undertaken: a check of the oil boxes above the running plate. I did not know then, and still do not know, what part of the motion these boxes served, only that the oil was conveyed to it down tubes by strands of bright green worsted. My brother and I were given a length each to keep. No Garter Knight, on the day of his investiture, could have matched the pride we felt at that moment. The driver then backed the locomotive on to the train. Buffers kissed and, to our alarm, Mr Loveday jumped down between the engine and the first coach to connect the screw coupling and vacuum brake and afterwards some cables on either side.

We stood back on the platform while all this was going on, then the time came to mount the footplate. We were helped up into the cab, a surprisingly roomy but cosy place and, being that of a tank engine, enclosed from the back as well as the front. The back head of the boiler projected a little into the cab and attached to it were the two glass columns which Mr Loveday used to check that it was time to turn off the injectors, the steam-driven devices which force water from the tanks into the boiler. A huge lever, the steam regulator, and a pressure gauge completed what he grandly called 'the boiler fittings' and to either side were the reversing and brake levers. A look up to the pressure gauge, the single hand of which was about to cross a line permanently inscribed on the dial, 'Time

for her breakfast, boys', and Mr Loveday opened the fire-hole door
ready to build up the fire and, at the same time, cool it, perhaps to
avoid the engine blowing off in the station.

We had seen fires before, of course, and had indeed been
responsible for a number of spectacular ones that later had to be
put out in rather a hurry. Nothing, though, had prepared us for the
view through the open fire door. Coal and flames were coalesced
into a single roaring mass, the heat from which forced us to the
back of the cab. Unwittingly we were in the way and, before Mr
Loveday could add fresh coal, the safety-valve lifted. What a fright;
we hardly heard the guard's whistle from the other end of the
train, but Mr Loveday nodded to my father and my brother was
held up high enough to grasp the ring at the end of the chain
controlling the whistle. He got two pulls at it before the driver
attended to the reversing lever and, after Mr Loveday had released
the brakes on both the locomotive and the train, opened the regulator. Much hissing from in front as clouds of steam poured from
the open draincocks of the cylinders and then, at last, the little
engine began to speak to us in the soft, deep-throated voice
we had only heard before from the line-side. We glided away from
the platform, Mr Loveday leaned out of the cab to take a token
from the signalman – we were on a single line – and we were on
our way.

We were too small to look out of the spectacle plates so were sat
on the end of the right-side tank projecting back into the cab. We
had a good view from here as the surprisingly lively little engine
whirled us up the gradients out of Chesham, along the embank-
ment past the watercress beds and through the cuttings and
woodlands of the Chess valley. There was so much to see both

inside and outside the cab that we occupied so proudly. Both Mr Loveday and the driver had lemonade bottles full of cold tea, complete with milk and sugar. I was given some out of an old enamel mug taken from a little shelf at the back of the cab. It was very refreshing and I have had a soft spot for cold tea ever since.

Eventually, to our great disappointment, we reached the junction with the main line at Chalfont. We were lifted down out of the cab and walked forward to admire the engine. We would really have preferred one with outside cylinders. This one's were down between the frames and we were not tall enough to get a good look at them and their valve gear. A little trickle of hot water was running out from under the cover of the steam dome and, just as we were studying this, the safety-valves lifted again and our fingers went back into our ears. It was time to leave but, this time, not on the footplate. We were taken back to the guard's van, at the end of which three large windows looked straight down the track. The driver joined us but Mr Loveday remained on the footplate. The Chesham Shuttle was an auto train, an ingenious arrangement whereby, on the return journey, the engine pushed the train and the driver controlled it from the guard's van. In compensation for my not getting to pull the whistle cord, I was given responsibility for a hooter operated by a button and designed to warn gangers along the track and motor cars at level crossings. They were well warned that day.

It might seem odd that Peter's and my interests in the natural world should have proceeded in parallel with our obsession with steam locomotives, but the fact remains that they did. Perhaps it was because they seemed to us to be frighteningly alive. As Peter and I stood on the station platform, fascinated but a little fearful of the hot hissing monster before us, it is tempting to imagine that our distant ancestors felt a similar thrill in the presence of a mammoth. Indeed, the parallel between a locomotive in steam and a large herbivorous mammal, be it a mammoth or an elephant, is not entirely superficial. Both derive their energy through the oxidation of plant material, fresh in the case of the mammal and fossil in the case of the locomotive. Both in turn are ultimately dependent upon solar energy stored through the entropy-defying process of photosynthesis, the one by contemporary plants, the other by giant ferns and horsetails that last swayed above the warm swamps of the carboniferous era some 300 million years ago.

The final development of our steam obsession came with the realization that it was possible to take a 'bus ride' to Berkhamsted station on the West Coast main line. This was the territory of the old London, Midland and Scottish Railway, a forward-looking organization in which, unlike our first love, the LNER, successive locomotive engineers had pursued policies of standardization. The result was that members of the more modern classes of train shared many features in common and, by looking at some of the older examples, it was possible to see the first expression of characteristics which were later to spread to whole classes.

As my brother and I stepped back awestruck at the thunderous impression created by the mighty, 160-ton 'Duchess' class Pacific at the head of the *Royal Scot* or marvelled at the loose-limbed elegance of the Midland Compound piloting the semi-fast to Crewe, we were unconsciously learning to compare form and function in a critical way – skills which later were to prove so important in our professional lives as biologists. The result of conscious design, the development of modern steam locomotives from their

Stephensonian common ancestor was nevertheless a process in which selection against types which failed was as important as technical innovation. To this limited extent, there were parallels with the unconscious but ruthless process by which natural selection acting on variation among individual organisms has been the powerhouse for biological evolution. But we knew nothing of this as we cowered from the edge of the platform when the *Royal Scot* rushed past.

BIGGER FISH TO FRY

For a few years after the war, my mother and father were prosperous enough to send my brother and me to kindergarten, in our eyes a highly unwelcome interruption to our lives. The little school, which was run by enormously tall nuns, was nowhere near the Chess and the fish we had become so keen to observe and to catch. Our next school, however, was opposite the river, although its pupils were quite different from those at the Catholic kindergarten. They were extremely friendly, yet they had not led sheltered lives like us. I shall never forget the shock I felt when, at our first Christmas there, one pretty little girl expressed the familiar carol in the words, 'While shepherds watched their turnip tops all boiling in a pot, the angel of the Lord came down and scoffed the bloody lot'. Another girl was astonished that I had never heard of chewing-gum. She extracted a greyish ball from a fold of her greyish handkerchief and popped it into my mouth. This loss of innocence took place by some railings through which I saw a tiddler flash (the rows of guanine crystals no longer being in line with the incident light) as it turned in the river below, and it was not long before Peter and I were after the sticklebacks again.

We discovered that there were two sorts: the three-spined *Gasterosteus aculeatus* L. with its colourful males and what we then called the ten-spined – now correctly called nine-spined – *Pungitius pungitius* (L.). The ten-spined fish were not so common and we usually caught them closer in to the edge and in weed. In due course, we were to learn that both species of stickleback are related to pipe-fishes and sea horses and that the brackets round the '(L.)' of *P. pungitius* were not a typesetting error but an indication that its generic name had been altered since Carl Linné (his name is usually Latinized as Linnaeus) had first classified it.

Our spirited cousin, Stuart, lived at Chorleywood, within walking distance of one of the lower reaches of the Chess. Here the river is no longer a brook but a fully-fledged chalk stream and it contained species that we had only ever read about in our much-thumbed copy of *The Observer's Book of British Fishes*. Stuart had been spared the years of sheltered gentility inflicted by nuns whose

The Observer's Book of Freshwater Fishes of the British Isles, *a much-thumbed treasure that sparked a lifelong interest*

only contact with the natural world appeared to be the bunches of damp shamrock with which we were issued on St Patrick's Day. While we were praying to St Teresa or St Anthony to help Sister find her board rubber, Stuart was out with his pals catching smooth newts and frog tadpoles in the ponds on Chorleywood Common and perfecting the netting techniques that we were soon to put to work in the strongly flowing waters of the river. Without our realizing it at the time, our training as biologists had begun. The bed of the Chess at Chorleywood is a mix of flint gravel and sand with silt both at the edges and entrapped in the roots of great swirling fronds of water crowfoot in mid-stream. We soon found that wellingtons were an encumbrance and that the key to understanding the river lay in being completely amphibious.

We discard our shoes, the hated buckled leather sandals which, by now, we consider sissy, and step into the brilliant clarity. It is late spring and the river feels as if it is made of ice. For a few seconds our minds are blanked out by the pain. Then the blood vessels in the skin of our legs contract, temperature receptors cease to respond and we find that we can function again – until, that is, we venture into deeper, more strongly flowing water. Here the horrible sensations are repeated further up our short legs.

Stuart is already well out in the river. He has the best net, one without any holes and with a square frame that he can press hard down on to the gravel. We see him turning his net inside out into a large paint tin and then moving on, sweeping one foot down through the water crowfoot and jamming his net against the weed. Our own efforts are less assured. The gravel pebbles are covered by a scratchy, chalky deposit and every so often a sharp broken flint assaults the soft skin of our feet. I come across a much larger flint and lift it gently aside. A mist of disturbed silt clouds my view but, as it clears, I see a small fish, trustingly still on the gravel. I attempt to net it but it gets past the curved frame of the net and all I find in the folds is a twig with legs – a caddis larva – and some grit. What a disappointment! The silt clears again, a jerky movement attracts me and I realize that, miraculously, the little fish is still there. Gently I push the net behind it and two more sweeps of its large pectoral fins carry the fish forward. But I can still see it and, scooping grit and fish together, I secure it.

My brother is similarly successful under another stone and, by the time our cousin comes ashore with his paint can, our catch is into double figures. Proudly we show them off. How contemptuously he dismisses them: 'Only miller's thumbs. We don't need any more of those.' We are not so sure. We tip them out on to the short grass by the wooden foot-bridge. There is one large one, nearly six inches long, two or three are medium-sized and the rest are all small. Without knowing it, we have representatives of three different year-classes of the common bullhead, *Cottus gobio* L. Years later, I encounter its close marine relative *Myoxocephalus scorpius* (L.) in the creels of Scottish lobster fishermen. The books call it the short-spined sea scorpion or father lasher, but the fishermen call it the 'Jimmy Cunter'.

We decide to keep the largest specimen and release the rest near the bridge. (It would have been better to have put them back near the stones where we found them, but at least they were downstream of their shelter stones and if, like salmon and trout, bullheads use olfactory cues to find their way around, they had some chance of a safe homecoming.) Now it is Stuart's turn. No bullhead lowers the tone in his paint can. Half a dozen minnows, *Phoxinus phoxinus* (L.), dash about at the surface and one jumps out on to the grass, pale gold below and patchy brown above. The minnows – some three to four inches long – probably come from the same shoal of pre-spawning adults that is now patrolling the run between the water crowfoot beds. Two stone loach, *Barbatula barbatula* (L.), lie long – over five inches – and orangey brown at the bottom of the can. We tip out the larger of the two and admire it grudgingly. Six barbels hang from its mouth. It starts to squirm about and we nearly lose it among the stems of last year's reeds. The loach and the minnows are to occupy a large tank in Stuart's bedroom. Both species do well, the minnows active in mid-water and the loach snuffling like springer spaniels for the chopped earthworms we drop in. At other times they seem to lie quietly on the sand until changes of air pressure cause them to dash to and from the surface.

From that summer onwards, the Chess was to be our teacher. Unreliable home-made nets gave way to flat-framed shrimping nets purchased on holiday. Paint tins gave way to plastic buckets and we begged a white developing dish from my father so that we could sort our captures properly. For the first time, we were able to take a proper look at some of the animals upon which the fish depended. There were the water-snails that scraped a living from the surfaces of the starwort leaves by rasping them with the abrasive rust-red tongues biologists call radulae. In the same sample squirmed the flattened grey-green leeches that, with the brown trout, were the snails' principal enemies. On the bottom of the dish, caddis-fly larvae dragged their clumsy cases of sand and twig. What a contrast with the delicacy of the mayfly nymphs, now tiptoeing alongside them, now swimming uncertainly, flexing their three fragile-looking tail whisks like the flukes of minute dolphins. Here was our first glimpse of the strange world of the fishes whose lives we were so anxious to share.

THE LITTLE NIPPER

If the age of steam was able to reawaken cerebral circuitry hard-wired into Man since the age of the mammoth, so the invention of the gun represented a further expression of Man's ability, which he shares with few other predators, to strike his prey at a distance. Bows and arrows, catapults and, above all, guns of all kinds are of the greatest interest to boys, especially when that interest is encouraged by sympathetic adults.

My paternal grandfather was comfortably off in a small way. He had a De Dion-Bouton motor car but he did not have the resources to keep horses, a *sine qua non* for those with ambitions to join the local yeomanry regiment, the Royal Bucks Hussars. He was content to join the Bucks Volunteers as a rifleman, which he duly did some five years before the outbreak of the Great War, and as a result he became an expert marksman. (The value of such training was to be amply demonstrated by the British Expeditionary Force at Mons in 1914, when a much larger German force was set back on its heels by the rapid and accurate fire of the British Lee Enfields.) My grandfather volunteered for service in France on the outbreak of war but, to his great disappointment, was unable to conceal his severely damaged heart valves, the effects of rheumatic fever, from the army doctor and was rejected out of hand. His interest in rifle-shooting remained undiminished, however, and when my father began to show an interest in sporting pursuits, he received every encouragement both from his father and from his maternal uncle, John, who farmed a few fields near Ford in Buckinghamshire.

Like his uncle, my father's passion was that of the hunter rather than of the soldier. He and a boyhood companion from a nearby farm became expert shots with a catapult. They had learned that the first step in making a good catapult is to find a naturally growing vee-shape in the hedgerow and then to bind the two shoots together so that, by the time the body of the catapult is cut out of the tree some weeks later, its two arms are more or less parallel. One end of each elastic, which should not be too strong or accuracy suffers, is held in a cleft at the end of each arm. The pouch is best made of soft leather. I still have my father's catapult, which was made of varnished apple wood. The clefts have been hardened by gentle charring and the end of the handle is bound with a single strand of polished copper wire to prevent splitting. It is a princely weapon.

House sparrows were the boys' usual quarry and they had found by experience that the best load was number-four-sized lead shot. On one dreadful occasion, a larger fowl was added inadvertently to the bag. Just as they walked into the farmyard, catapults in hand, a very aggressive cockerel, one well-known to the boys and whose raking spurs they had learned to respect, advanced to the attack. Thinking to scare it off, my father's companion casually picked up a pebble and engaged the cockerel with his catapult. By an incredible fluke, the poor bird's beak was driven backward into its skull and it fell, spouting blood, in a flurry of feathers. Fortunately, there were no witnesses and the boys were never suspected. My father

always insisted that one should never shoot what one was not prepared to eat, and so no doubt the cockerel was discreetly plucked
and gently stewed at some distance from the scene of the crime.

My father's first successful gun was a .22 smooth-bore saloon
pistol which fired rimfire cartridges loaded with small shot and also
what were euphemistically marketed as CB caps. CB stood for
conical ball and there was quite enough powder in the cap to cause
severe injury. In those days, of course, there were no draconian
firearms laws, and those few boys who were lucky enough to have
access to such exciting items as saloon pistols tended to be from
country families in which the safe handling of guns of all sorts was
taught from early childhood. My father's pistol bore the legend
'The Little Nipper' and so it proved. If the catapult could produce
enough sparrows for individual roasting over a fire, the shot cartridges of the Nipper were the key to bags large enough for
incorporation into shortcrust pies and suet puddings.

My father in 1938 with his first goose
(a white front) and Henry Aldridge at
Wells-next-the-Sea

Unlike my father, my brother and I never really made anything of our catapults. We were much influenced by the *Badminton Library* volume on archery which was full of illustrations of every possible sort of bow and contained a detailed account of the prodigious range and penetration of the English longbow. We scoured the hedges for hazel sticks straight and flexible enough to make into bows and strung them with whatever we could scrounge from our grandmother's kitchen drawer. For arrows we usually used the dead stems of golden rod, occasionally tipped with heads of slate and, very occasionally, dipped into the supposedly deadly poison present in the crushed red berries of lords and ladies (wild arum). We had the greatest difficulty in hitting even the largest of inanimate targets and never posed the slightest danger to small game of any kind.

Once again, our indulgent father came to the rescue. We had got up rather late after a disturbed night. Our fear of foxes had passed but we were prey to a new terror: the possibility that the house to which the family had moved was haunted. The basis for this irrational belief was undoubtedly the eerie flickering light with which the newly purchased paraffin night-lamp bathed our bedroom. On this particular morning, my father had left very early but returned before breakfast. We heard the gate latch and rushed to the bedroom window; small boys do most things at the double. My father paused and looked up and, from his pockets, slowly drew the butts of two Webley air pistols. We rushed downstairs and were sent straight back up again to get washed and dressed.

The new treasures were lying side by side on the kitchen table, a .177 Junior model for my brother and a .22 Mark I for me. Both were early versions of these famous marques, the Mark I having wooden grips and a side safety-catch with an engraved arrow on it. They were quite difficult to cock and, although intrinsically accurate, were much more difficult to use than the air rifles to which we graduated some years later. Nevertheless, it was not long before the wood smoke of our cooking fires was curling skyward, plump sparrows roasting on the sharpened sticks we had cut from the hedge. We were savage innocents living before Sir David Maxwell Fyfe had abolished rationing, and at a time when net-caught small birds had not long ceased to be part of the winter diet of poor country families. We learned the rudiments of cookery in the most direct way possible and the even more important lesson that taking life must never be done lightly.

Long before we were ourselves old enough to shoot, Peter and I rejoiced in the opportunity to accompany our father into the shooting field, to see him put his most recently acquired treasure through its paces. At first, it was all a little frightening. The report, even of a game gun, can leave sensitive young ears ringing; often it became a race between the gun snapping smoothly to my father's shoulder and the reflex jamming of our index fingers into our ears. Such was his natural skill that fear of the bang was soon forgotten in our rush to retrieve the grey partridges and occasional pheasants that thronged the little fields and beech hangers of Buckinghamshire in the years immediately after the Second World War.

Exciting as all this was, and however much we enjoyed the heady scent that lingers in a newly fired cartridge case, the greatest thrill was always to see one of the muzzle-loaders in action. Fine black powder, the flinty grains shining as they trickled from the measure at the head of the tooled copper flask, would be the first to be poured down the barrel. A gentle tap to settle the charge and then the wadding would be rammed hard down so that, on ignition, the powder would be able to exert its full power. Next the shot, again the butt tapped against the toe and lighter wadding rammed on top, gently this time to avoid deforming the lead pellets. Finally, with the hammer at half-cock, a percussion cap would be placed on the nipple and everything was ready. All that was necessary now was to pull the hammer back to full-cock and pull the trigger. The result was always so earth-shattering that, as far as Peter and I were concerned, whether or not anything fell to the shot was completely immaterial. An orange flash, accompanied by a crackling boom, would shoot from the muzzle and a dense cloud of smoke, smelling strongly of bad eggs, would hang in the air. If any one event was to set us on the road to becoming wildfowlers like our father, this was it.

Most of my father's success as a wildfowler was achieved before the war at Wells-next-the-Sea in Norfolk, where his host was Henry Aldridge, a retired RAF officer whose flying career was cut short in a crash landing in which he lost a leg. After the war, the opportunities for wildfowling at Wells were much reduced. None the less, he persevered and acquired an eight-bore by the premier London gun-maker James Purdey. It was a most magnificent gun, a larger version of the back-action hammer game guns favoured by such great shots as the Marquess of Ripon and His late Majesty King George V. But the numbers of grey geese frequenting Wells had fallen so dramatically that opportunities for using the Purdey were few. Its last outing was in pursuit of a large white greylag much sought after by the local gunners. My father had the good fortune to fall in with it, and the Purdey did its deadly work. My brother and I had gone to bed when my father returned and we awoke to find him standing in the doorway of our bedroom with the Purdey in his right hand and the white goose and a redshank in his left.

With the departure of the geese from Wells, my father's interest in the Purdey waned and it was not long before the great gun was on the market. I last saw it as a child on top of its open case, its beautifully engraved lock plates and action catching the morning light. My father sold it for £45. Many years later, I saw it again in the catalogue of a London auction house with a guide price of many thousands. From that moment onwards, however, I was determined to obtain an eight-bore of my own.

If the first inklings of my later interest in the evolutionary relationships between living organisms was to be stimulated by comparing the designs of steam locomotives, it was strongly reinforced by looking with an increasingly informed eye at the succession of nineteenth- and early-twentieth-century guns that passed through my father's gunroom. From his six-bore muzzle-loading pigeon gun to his favourite sixteen-bore game gun by the famous London maker Charles Lancaster, my father's guns were snapshots drawn, like fossils, from a period of rapid technical innovation which began at a time of stove-pipe hats and powder flasks and ended just before the First World War with guns indistinguishable in design from those used in the present day.

As I had already noticed with steam locomotives, at each stage of the development of guns – from flint to percussion, from pin fire to centre fire and from hammer gun to hammerless ejector – new external features were acquired and older ones lost. But again, as with locomotives, not all of the old structures were entirely discarded. Some were retained as decorative reminders of earlier times in the development of the sporting gun. It was to prove only a small – but for both Peter and me, highly significant – step to see how such relict structures have also been retained during the evolution of fishes.

———

RED FLOATS AND ROACH

———

Our obsession, for that is what it became, with fish we owed to our maternal grandfather. He had been a keen footballer before the First World War and both he and the whole of his team volunteered for service in the first wave of enthusiasm in the autumn of 1914. He joined the Bucks Battalion of the county regiment, the Oxford and Bucks Light Infantry, an elite corps which prided itself on being 'the fastest-marching regiment in the British Army', and was later transferred to the Machine Gun Corps. He was lucky enough to find himself in the army commanded by General Sir Herbert Plumer, a forward-looking officer who was to adapt his tactics to the needs of trench warfare more effectively than any of his contemporaries and whose professionalism saved many lives on the Western Front. Later, my grandfather left the Western Front with a British Army that was sent to stiffen Italian resistance, then faltering in the face of the armies of the ramshackle Austro-Hungarian Empire, an alliance which the Kaiser famously compared with being 'shackled to a corpse'.

Although a shellburst had nearly accounted for him on the Western Front, it was the sight of so many dead men and horses on the Italian front that had left the most lasting impression on my grandfather. Terrible as these events were, all were overshadowed for him by the loss of his beautiful wife, Edith Dudley, who died in 1917 of what was then called childbed fever. Some years later he was to marry her plain but wonderfully kind sister, Mary, whom I was to know as my maternal grandmother but who was really my great-aunt.

By the time the war ended, my grandfather's footballing days were also over. His spiritual equilibrium was restored by the winter days he spent with his older brother, fishing for jack (small pike), *Esox lucius* L., in the Aylesbury arm of the Grand Union Canal. In the summer, he would take his whole family there to fish for bream, *Abramis brama* (L.), roach, *Rutilus rutilus* (L.), and perch, *Perca fluviatilis* L. My brother and I were to join these summer expeditions after the Second World War, watching my grandfather's little red float and, later, our own, bob and then disappear as bread paste or worm was taken.

At first, the waters of the canal were greened by the swarming cells of algae but, as the summer progressed, the nutrient salts in the canal were exhausted and the water began to clear. By the holidays, the canal was as bright to the eye as our beloved Chess. Now, in the full light of the long summer days, the stands of starwort and water crowfoot grew quickly, using up the remaining dissolved phosphate and nitrate so thoroughly that the algae were not able to cloud the water until the late autumn.

'Stop fidgeting, you boys,' urges my grandfather. We do our best to keep still and, to our delight, a shoal of minute roach emerges from the shadow cast by the sallow bushes along the far bank. They are fish of the year, risking all to feed and grow before the curtain comes down on the warmth and abundance of summer. How close they are to the surface, keeping together with the stiff movements of bodies that seem only to flex from their dorsal fins to their tails. What a contrast with the sinuous swimming of the trout fry we had seen holding station above the bed of the Chess. A little splash, and one of the roach fry dashes at a fallen moth caterpillar that has blown off a hawthorn bush. Peter and I can contain ourselves no longer. We stand up and the shoal immediately disappears.

'For goodness sake, keep still!' But we need no telling; for a time at least we have learned our lesson, and it is not long before our patience is rewarded as a small school of perch emerges from a crowfoot clump. They stay in mid-water and, as they come into the sunlight, we see the dark vertical stripes that help to break up their outline. They are gone before we see any of them feeding but, with their departure from the stage, ghost-like roach, large enough to have been the parents of the innocents sporting at the surface, glide stiffly into view. My grandfather signals to us to stay still in the hope that these bigger game will show an interest in the bread paste suspended below his float. One does, and the float gives a little knock. For Peter and me the excitement is all too much. 'Look at that!' Peter has the sense to keep still, but again I stand up and the roach

The perch, Perca fluviatilis *L., a fervent biter highly regarded by small boys*

dart into the shadows. My grandfather, the kindest and most patient of men, says nothing.

The sequence in which we boys had seen first the roach fry, then the perch and finally the large roach made perfect biological sense. The smallest fishes are vulnerable to such a large range of predators – everything from tiny corsairs like the larvae of dragonflies and beetles to perch, pike and gulls – that leaving shelter to feed and grow and thereby to shorten the list of potential assailants, is a risk well worth taking. The balance shifts as the fish grow bigger and increase their investment in the fat and protein they will ultimately need to reproduce. Hence the larger roach are also the shyer ones. They are essentially defenceless fishes which depend for their survival on their ability to detect and flee from danger. Their skin is so soft that even a touch from a dry hand is enough to tear it open and displace the flat cycloid scales beneath. What a contrast with perch! Protected by robust ctenoid scales that project through their tough skin to form a light armour analogous to chain-mail and able to erect a spiny dorsal fin which renders them difficult for birds and pike to swallow, they can afford to take chances that no roach would countenance.

So shallow had the old canal become, and so clear its water, that even the deepest fishes we saw had probably been frightened because the abrupt passage of our brightly clad bodies had stimulated movement receptors in their retinae. It may not have been the only way in which we had inadvertently communicated our excited presence to the fish. The surface of water is an even better reflector of sound than it is of light, so the perch and roach were probably unaware of our exclamations. However, the lower-frequency disturbances created by the movements of our feet, and readily transmitted through the bank and into the canal, were another matter altogether. Water is virtually incompressible and transmits sound wonderfully well and at five times the rate of air. A sound wave is a transient increase in pressure in which the particles of the medium through which it passes also move. Particle movements are greatest close to the source of the sound, and fishes have specialized organs called neuromasts on their heads, and in the open tube that forms their lateral line, which are able to detect these so-called near-field particle movements directly. To pick up sound farther away, they rely on the fact that their bodies are comparatively transparent to sound waves, but the dense bones called otoliths, three on each side of their heads, are not. As the sound wave passes through the fish, so the otoliths lag behind. Because each is suspended in a jelly into which modified neuromast organs project, the minute relative movement between each otolith and the rest of the fish is signalled to the brain.

It is not the only trick up their sleeves. Perch and roach are able to maintain near-neutral buoyancy because they possess gas-filled swim bladders. When a sound wave meets a swim bladder, it is briefly compressed and so acts as a secondary source of particle displacement which the fish can detect. Particularly is this true of the timid and unprotected roach which, like all members of its group, has bones called Weberian ossicles which articulate between the head of the swim bladder and the inner ear.

As Peter and I grew older, so the excitements of extreme youth gave way to the strengthening imperative of the hunter. Although our angling success rarely approached that of our grandfather when fishing alone, and although we never caught a fish greater in weight than a pound, the blank days became less frequent. If our catches were small, unlike modern coarse fishermen, we nevertheless ate everything we caught. We quickly became aware that freshwater fish which live in muddy habitats acquire a curious earthy taint. It is caused by a substance called geosmin produced by a yeast-like fungus that lives on the surface of mud. My grandfather showed us how to get rid of it by soaking the cleaned and scaled fish overnight in salted water, to which a little vinegar had been added. We used to eat the fish for breakfast, tossed in flour and cooked in dripping over a brisk heat.

ORDOVICIAN ECHOES

Peter and I are back at the Chess, older now and masters of a wide enough range of catching techniques to sample all of the species present in that part of the river. We have with us our youngest brother, Graham. It seems such a long time ago that we first saw him in his carry-cot. We had been staying for some time at our paternal grandmother's catching the large and yellowish frogs which, after an afternoon's handling, would mysteriously expire. We had only the haziest idea of where Graham had come from. We knew that we ourselves had been 'made by God' but that, in some strange way, my mother had obtained our new brother from a hospital. (In due course we found out that he had been born at home and that my mother's visit to the hospital was for a post-natal check-up.) We gave the problem no further thought and it was to be some years before an honorary aunt startlingly revealed that babies 'come out of people's bottoms'.

We have ridden over by bicycle, the pace set by Graham. It was possible to free-wheel some of the way and my mind drifted to a film I had seen at school about the 'great liners' – not the Cunard giants but the stout vessels from north-east Scotland that fished for halibut using 'great lines'. I was no longer grasping the handle-bars of my Rudge bicycle but the wheel of a fishing boat as it met the Atlantic swell. Peter and Graham had become my fellow skippers and the Chess the deep-water banks where we were to shoot our lines.

How ridiculous: I tell no one about my day-dreaming but, gripped by the strange insecurities of early adolescence, wonder if there is something wrong with me. We arrive at the river and discard our bicycles. We are comprehensively equipped with nets and rods tied to crossbars, and open-topped thermos flasks in which to bring back smaller specimens for our aquarium. While Peter and Graham search for crayfish, *Austropotomius pallipes* (Lereboullet), under the largest stones, I set up a miniature rod and attempt to 'long-trot' for minnows between the beds of water crowfoot. I have a size eighteen hook which I bait with a single blowfly maggot. It drifts past the minnow shoal, every member of which studiously ignores it. I persevere, but with the same result. Surely they can see it? I think to myself. Maybe they can, but their central nervous circuitry will not permit them to put their lives at risk by breaking away from the shoal. 'Try it lower!' yells Peter. I push the little float up the line. The result is miraculous. The minnows compete to seize the maggot as it trundles through the shoal. By the end of the morning, seventy-two of the largest specimens lie on the bank under a damp covering of aquatic moss. After cleaning, and an overnight soak to remove any geosmin, they are to reappear, crisply fried, at breakfast.

While absorbed in the long-trotting, I glimpse a large gravid perch being swept down past the weeds over by the bank. It is alive but has lost control of its buoyancy. Perch are advanced, physoclistous fishes, unable to release swim-bladder gas into the foregut. Perhaps this one has been brought to the surface from the bottom of Latimer Lake by an angler and the expanded gas has yet to be reabsorbed into the blood. More probably, high levels of the cortico-steroid hormones associated with its sexual maturity have suppressed its immune system and an infection is drawing its life to a close. When I get back with my minnows, I find that my brothers have gathered the perch into a little backwater. It is still breathing and we have no desire to eat a gravid fish, so we push it out into the stream in the hope that, somehow, it might survive.

Peter and Graham have done well among the crayfish, close relatives of the common lobster, *Homarus gammarus* L., later to loom large in my professional life. They are a rich chestnut brown above, shading to cream with a hint of turquoise below. Both colours are cryptic. Seen from above, the crayfish's livery is a good match for the rusty red of many of the flint pebbles; and the underside, which a fish would see when the crayfish is attempting to escape in mid-water, is a fair match for the sky. Like perch, they enjoy the benefits of mechanical protection, but instead of the perch's flexible chain-mail, crayfish use the equivalent of plate armour and, in their powerful claws, possess formidable weapons of their own.

The armour of crayfish is also their skeleton, a form of construction known to motor-car manufacturers as monocoque. In many ways, it is a more efficient approach to providing the rigidity and flexibility required in an organism which depends on compression struts to walk, swim, feed and fight. However, a rigid external skeleton has one overwhelming disadvantage. It cannot expand to accommodate growth, so each time the crayfish increases in size, it has first to moult and then wait for some hours for its new, larger skeleton to harden. Throughout this period, it is highly vulnerable to attack by fishes, birds, aquatic mammals and other crayfish, so finding adequate shelter is essential.

Although crayfish die when put into sea water, their immediate antecedents were marine. The young of these ancestors lived in the plankton for several weeks and looked very unlike the adults into which they would eventually metamorphose. But for animals living in a strongly flowing river, a long planktonic larval stage would be highly disadvantageous. In the crayfish, therefore, these stages of development take place within the eggshell. The young animal which hatches from the eggs, held by appendages called pleopods on the underside of the female's tail, strongly resembles the adult.

We avoid the eye-watering pressure of the crayfishes' claws but get the odd squeeze as a flapping tail traps a finger. We return the little ones to the river. They pulse away backwards, their claws trailing. Some of the large females are carrying eggs and I would like to be able to say that these mobile maternity wards joined the small specimens back in the river. The thought, however, never even entered our heads. We put damp moss into the bottom of an old gas-mask bag and gently lay the precious crayfish on top of it, layering moss and crayfish until the bag is three-quarters full. We fill it up with watercress to eat with the crayfish when we get home. We have not yet learned that preliminary immersion in warm water has an anaesthetizing effect on aquatic crustaceans and our method of cooking the crayfish is to plunge them straight into boiling salted water. The quality of mercy is but poorly developed in boys.

Lunch-time is still an hour away. Peter shows Graham the small trout that can usually be coaxed into the net from their shelters in the undercut banks and then goes off on his own in search of greater fish. Jamming his shrimp net against the side of the water crowfoot, he works his way up one of the larger beds, sweeping his left foot down through the weed to herd any sheltering fish into the net. I am keeping an eye on Graham when I hear Peter shout, 'I've got one!' He certainly has. He scrambles to the bank, wetting himself to the waist with his splashing. He holds his net high and, in the bottom, the equivalent of the trawlerman's 'cod end', a magnificent wild brown trout lunges from side to side.

The Chess is a highly productive stream. Its main source of energy, sunlight, readily penetrates through its often gin-clear waters and both algae and rooted weeds do well. Salts from the aquifers, the underground reservoirs in the chalk which feed the river with naturally filtered water, ensure that the development of no plant or animal is inhibited by mineral deficiency. Worms, molluscs and insects thrive and the fish that feed on them grow fast. Trout of the size that my brother had just caught take minnows and bullheads as well as smaller prey and live at the top of the Chess food web. One of the other reasons that these fish do so well is that only so many are invited to the banquet.

Trout need clean gravel in which to spawn so that their eggs can spend the winter aerated by water flowing through it. Female trout deposit their eggs in depressions which they create in the gravel by turning on their sides and working their tails. This activity continues upstream of the deposited eggs which are then safely buried by the displaced gravel. In rivers like the Chess, the surface of the gravel pebbles is often roughened by mineral deposits which accumulate around the cells of primitive photosynthetic organisms called Cyanobacteria. At best, these rough deposits, which were so testing to the soles of our feet, can make the gravel harder for the trout to displace while at worst, and in combination with silt, they can lock the bed solid. The result is that the opportunity for successful spawning, and good egg survival over the winter, are restricted and the numbers of young trout with it. Narrow is the gate by which a Chess trout enters its chalky Eden.

But we know nothing of the dynamics of trout populations. All we know is that Peter has just caught a large trout and that it is now lunch-time. It is just as well, given that Peter is wet to the waist, that I have been entrusted with the matches. We raid the hawthorn hedge running down to the river for twigs, snapping them from dead branches still upright, rather than picking them up from the ground where unseen dampness can make them difficult to light. Dry twigs, a warm day and the paper from the sandwiches we have been given, ensure that the 'one-match' light-up on which we pride ourselves (we had heard, I suspect quite wrongly, that Boy Scouts needed two) is easily achieved. We have to cut off the trout's head to get it into the pan. The muscle cells that make up the body of the fish contract strongly as it hits the smoking butter and we have to turn it over quickly to keep it straight. It tastes fine, provided we ignore the pinkish liquid, with something of Good Friday about it, that has soaked into the slices of bread on which we have laid the fish. There is no hint of geosmin taint, probably because the trout has been feeding mainly on wind-blown insects and mayfly larvae rather than caddis and water-snails.

That afternoon we decided to explore a section of the river a little upstream of our usual fishing grounds. We were on the threshold of making the greatest discovery of our lives as amateur naturalists. When we reached the lower end of the stretch we planned to explore, we noticed that the gravel moved about more under our feet, and that there was less of a visibility problem from disturbed silt. We were not the only vertebrates, back-boned animals, to have made this discovery. For once, I was in the right place at the right time. There in front of me were three slender forms lifting the stones with their sucker mouths. I had seen my first specimens of Planer's lamprey, *Lampetra planeri* Bloch, *Lampetra* meaning 'stone sucker'. They were very trusting, and with one sweep of my left foot and a forward thrust of my large shrimping net, I was able to secure all three.

'Quick! Come and look at this.' No response: Peter and Graham are too far downstream to hear me at first, so I wade in their direction and show them the treasures at the bottom of the net. Thank goodness we have tied up the holes in it. We dare not risk losing the lampreys in the bankside grass, so keep them in the net until we can release them into the white developing dish. They are battleship

Lampreys, including Planer's lamprey,
Lampetra planeri *Bloch*

grey above and a pearly white below. There are no paired fins but a single nostril, two bright eyes and seven gill openings. There are no jaws, only a permanently open sucker which one is using to hold on to the bottom of the dish. There is only one thing to be done. We part-fill a thermos flask with cold water from the river, grab the gas-mask case with our 'commercial catch' of minnows, crayfish and watercress and set off at once for home. We keep the lid off the thermos and drop in a twig which floats on the top and prevents slopping. When we get back, we check the temperature of the water in our display tank which already contains a roach, minnows, stone loach and a small crayfish. The tank is in the coolest part of the house but dipped fingers tell us that its temperature is still well above that of the water in the thermos. We beg a plastic bag off my mother, pour in the river water with the three reassuringly active lampreys and float the lot in the aquarium.

After an interval – probably far too short – to allow the lampreys to get used to the warmth of the tank, we up-end the plastic bag. The lampreys make for the bottom and by evening they are lifting stones just as we saw in the river. They are forgiving guests and we enjoy watching them for several days until an overnight fault with the aerator brings their lives, and those of the fish, to a premature end. The distinction is important: lampreys are not true fish but close relatives of a particular group of jawless vertebrates, with a fossil record going back nearly half a billion years. We cannot bear to throw the lamprey corpses out but transfer them to methylated spirit to join a bottled grass snake, lots of mammalian and bird skulls (including that of the white goose), the blown eggs of common birds, the stuffed skins of mice and a large collection of local fossils.

The next piece of the lamprey jigsaw was the discovery, by our cousin Stuart, of large numbers of their filter-feeding larvae burrowed in the beds of silt in the slack water near the bank of the Chess. The larvae of lampreys are known as ammocoetes. When they first hatch from the thousand or so eggs laid by a female Planer's lamprey, they are little more than half a centimetre long. They burrow in silt with their front ends sticking out into the river. They feed by filtering out very small organisms from the water column and retaining them, fly-paper like, on slime produced in their throats. By the end of their lives as larvae – a period which may last as long as seven years but is probably much shorter in productive streams like the Chess – the ammocoetes transform themselves into the free-living adults that we had found moving the stones. Unlike those of the river lamprey, *Lampetra fluviatilis* L., to which it is very closely related, the adults of Planer's lamprey are like mayflies in that they are purely a reproductive phase which does not feed. Whether they feed or not, all lampreys die after spawning. We were fascinated to learn these details and also that, until halfway through the nineteenth century, ammocoetes and lampreys were believed to be entirely separate animals.

As for the corpses in the methylated spirit, they were not allowed to rest in peace. I opened two of them up with a razor blade and found the gut to be thread-like and apparently empty. The discovery of the lampreys marked the end of my early life as an amateur naturalist. From that moment on, for reasons I did not then really understand, I was determined to make a career in biological science and, preferably, the science of fish and fisheries. Looking back at that Damascene moment, gazing at the silver-grey bodies busy among the stones, I have often wondered what made it such a decisive turning point. Part of the magic came from its unexpectedness. Here we were, in the very middle of England, surrounded by familiar plants and fishes, suddenly confronted with a scene that had probably first been played out several hundred million years ago, not long after the time when the fishes which had the potential to develop articulated jaws from their gill arches separated from those that had not. So important was this development that the jawless fishes are nowadays represented only by two unrelated eel-shaped forms: the sucker-mouthed lampreys and the strange hagfishes which attack their prey with bilateral nippers like those of many marine worms. Years later, off the coast of north-east England, hagfish, *Myxine glutinosa* L., were also to play a small part in my life as a natural historian.

SALT MARSH AND SEA

My father's interest in wildfowling had led him to form a strong attachment to Wells-next-the-Sea and the salt marshes of the north Norfolk coast. For many years, it was to be the scene of our seaside holidays in mid-summer. Then as now, Wells was a working port. There were railway sidings on the quay and the coastal traffic included Thames sailing barges, surprisingly large shallow-draft vessels which would nevertheless be worked by just two men. Our first visits coincided with the last gasps of the family's prosperity and we stayed at Chadd House with my father's wildfowling guide Henry Aldridge and his wife Ivy, who, as far as we could see, did all of the work.

Peter and the author (third from left) and two older chums find a very dead harbour seal

The 'next' of Wells-next-the-Sea refers to the fact that Wells and its harbour are situated at the head of a mile-long channel across from which lie great areas of salt marsh. In the summer the marshes are carpeted with the cornflower blue of sea lavender and scented by the soft fronds of ploughman's spikenard. Along the margins of the creeks, great strands of brilliant green samphire (pronounced 'sampha' in Norfolk) plants grow like miniature forests of cactus. Two memories remain from those earliest visits: the fascination of seeing an artificial leg for the first time – Aldridge's – and the first taste of cockles, *Cardium edule* (L.).

The cockles were taken from the same sand flats that extend along the coast to Stiffkey and produce the renowned Stewkey (an old name for Stiffkey) blues, believed by the locals to be the finest cockles in the world. We learned to leave them overnight in clean sea water to give them time to spit out any sand trapped inside their shells. They owe their wonderful flavour, as do many shellfish, to trace quantities of dimethylamine sulphide, which they take in with their planktonic food. We also learned that they taste best when they are steamed just to the point of opening and not boiled. Later I discovered that it is not safe to do this with cockles taken from anywhere near sewage outfalls. But we got away with it with our Stewkey blues, and there was no bacteriological risk with the samphire. The trick is to pull the plants up completely and rinse all the mud out of the white roots in the sea. You should boil or steam samphire for exactly seven minutes. You then hold each plant by the roots, dip it in peppered, melted butter and pull all of the green bits through the teeth. Eaten like this, it is superior to any other green vegetable and worthy of being served as a course on its own, accompanied by crisp rolls and a creamy New World chardonnay.

Compared with the Chess, net-fishing opportunities at Wells were rather limited. The brackish dykes or small creeks behind the sea wall were full of delicate shrimps and small fish which we later identified as sand gobies, *Pomatoschistus minutus* (Pallas). We also came upon three-spined sticklebacks, if anything larger than the ones often encountered in fresh water and with bony scutes (large scales) on either side. Occasionally, we would set a line for eels, *Anguilla anguilla* (L.) and once got a large one which fell off; from its size, it was almost certainly a female. We did little better in the channel on the other side of the sea bank but none the less added to our species list with the odd flounder, *Platichthys flesus* (L.), inshore sand eels, *Ammodytes tobianus* L., and once a magnificent great pipe-fish, *Syngnathus acus* L., well over a foot long.

Of all the fishes in British waters, none gives a more convincing impression of being an archaic throw-back to an earlier era than the great pipe-fish. Its long, slim body is encased in an armour of segmented plates, polygonal in section, from the back of the head to the vent and four-sided from there to the tiny tail fin. The great pipe-fish is hard and angular to the touch and crocodilian in colour, and when you first handle this languidly waving creature, the sensation is of having picked up a strange salt-water reptile, left over from the age of the dinosaurs. In fact, pipe-fishes are members of an advanced group of bony fishes which also includes the sticklebacks and the sea horses. Indeed, the great pipe-fish has the head of a sea horse and the tail of a stickleback. Altogether it is something of an exotic curiosity, threatening only to its tiny longshore prey of polychaete worms, crustaceans and fish larvae. As to its breeding habits, it is right up to date: the male rocks the cradle by incubating the eggs of his lover in a brood pouch of his own, from which he later launches the results of their passion into the sea.

Over much of the British coastline, especially in the north and west, land and sea meet abruptly with the crash of wave against rock. The separation is seen at its most extreme in places like the north-east of Scotland, where the farming and fishing communities live largely separate lives, marked by differences in speech patterns and even of religious belief. North Norfolk offers an opposite example. Sand dunes, salt marsh and fresh marsh, interwoven with grass-covered sea banks, soften the transition between the shallow waters of the Southern Bight of the North Sea and the rich acres which, in the hands of Thomas Coke of Holkham, were to pioneer England's agrarian revolution. This great hinterland was, and to an extent still is, the world of the marsh man whose main income might well come from work on the farm but could also encompass eel-fishing, wildfowling and cockle gathering. Other hardy souls would combine their lives ashore with such grimly physical occupations as mussel-fishing and bait digging.

Apart from the 'fowlers, the real aristocrats in our eyes were the whelk fishermen who represented the outermost fringe of the longshore tradition of north Norfolk and whose dangerous work often took them well out of sight of land. The whelk-fishing families had come to Wells from Sheringham back in the 1890s. Their boats were larger versions of the open, clinker-built craft used for crab-fishing off Cromer and were known as hubblers. They had a main mast forrard, a cuddy in the bows for storage and a little shelter, an inboard engine and a winch. They possessed the workmanlike beauty of so many traditional fishing vessels and we admired them greatly. Our special favourite, in the way of most boys, was the largest, the *William Edward*, a name we associated with kings of England but that was almost certainly a record of the names of the two brothers who had first owned the boat. Another of our favourites was called *Knot*, nothing to do with rope but recalling the plump, delicious wader of that name popular with 'fowlers in the days before wholesale bird protection. The smallest of the fleet was the *Tony*, a smart, largely black-and-white boat which had the sad distinction of losing one of its crew following an all-too-easy snarl-up with a 'fleet' of the heavy whelk pots.

The object of the fishery was the common whelk, *Buccinum undatum* L., a large, predatory sea snail which lives on other molluscs but which, like so many animals of the seabed, will also gorge on carrion. Like the periwinkles, *Littorina littorea* (L.), to which they are distantly related, they are sold to the public boiled and in the shell. There the resemblance ends. Unlike their smaller relative, whelks are large and chewy and, out of the shell, of most unprepossessing appearance. A large off-white foot gives way to a slimy topknot of brown and green internal organs. The secret of eating whelks is not to look at them first and not to breathe while chewing. Nevertheless, whelks have been prized in cockney London and in resorts like Southend for many years and it was this market, and no doubt related ones in Holland, Belgium and France, that were the main outlets for the fishery at Wells.

The returning whelk boats would appear on the flood tide beyond the bar which guards the entrance to the harbour channel, occasionally with their brown mainsails set. Sometimes they had mackerel, *Scomber scombrus* L., taken on feathered hand-lines and, if they had arrived at the bar too soon for the making tide to float them over it, holiday-makers would meet them there and go proudly home bearing damp parcels of the best eating fish in the sea, provided it is cooked simply and immediately over a brisk heat. Once in the channel, the hubblers would process the couple of miles or so past the lifeboat station and the sea wall to the east quay. With the tide behind them, the engines had only to idle to provide enough speed through the water for the rudder to do its work. Sally Festing, in her valuable archive *Fishermen*, describes the sound of returning whelk boats as 'grunting', and I cannot think of a better description.

Our particular favourites among the whelk fishermen were called Cox, one of the original families of Shannocks, a local nickname for Sheringham folk, who had come to Wells over sixty years before. Uniquely, they had exchanged their open clinker-built hubblers for ex-Royal National Lifeboat Institution boats which had come on to the market following their replacement by self-righting designs. Re-engined and fitted with a winch, they made fine whelk boats and, being fully decked with valved scuppers to cope with the sea's attempts to come aboard and join the crew, were potentially safer than the traditional design. The converted boats we knew best were called *Anne* and *Elizabeth*. They sat the sea like gulls, studies in grey and a blue which echoed their proud history as lifeboats. One of the highlights of our visits to Wells was watching *Anne* and *Elizabeth* unload their catch at the far end of the east quay, where the Coxes had their whelk-house. The whelks came ashore in bags loaded on to a little red flat-bed trailer powered by what sounded like a motor-cycle engine. When it was full, the little motor trolley would sputter up from the shore and into the whelk-house, a pantiled barn with plenty of hard standing for the bags of whelks and, at the far end, a boiling copper, set in stonework and into which successive netting bags of whelks were lowered. On bright days, the sun shining down from the roof lights would form shafts in the steam.

Presiding over all was the powerful figure of Mr Jimmy Cox himself, the most senior member of the family and of the Wells lifeboat crew. At his suggestion, we used to seek out the bright red hermit crabs, *Pagurus bernhardus* (L.), from among the steaming bags of whelks and crunch their claws between our teeth to get at the meat which to us tasted so much better than the chewy flesh of the whelks. The steam from the whelks was not the only contributor to the clinging whelk-house smell. In one of the cooler corners, there was a great, rancid heap of salt herring, no doubt much of it bought in from Yarmouth, which was then still the centre of the great East Anglian herring fishery. The herring was used to bait the small but heavy whelk pots. It was attractive to the whelks and, being some-what dehydrated by the salt, was tough enough not to break up in the tideway where the pots were fished.

•

The whelk boat Elizabeth *returning to Wells-next-the-Sea in 1957*

Going to sea aboard the *Elizabeth* was to be my first experience of the fishermen's world. My father and I arrive early at the east quay. Even so, there is no time to lose. The bait is aboard and the engine is ticking over. We are helped over the port gunwale and into a narrow space beside the engine cover, well out of the way of ropes and winch. Jack Cox lights his pipe, casts off forrard and gently opens the throttle of the six-cylinder Morris marine engine. The grunting gives way to a low rumbling vibration and we are off down the creek that leads to the harbour, where, against the wall, a small Dutch coaster is unloading fertilizer for the farms on Holkham Estate. We turn down the channel, the lifeboat station fine on our port bow and, over to starboard, acre upon acre of sea lavender. All around us terns are plunging after young sand eels and, over on the marsh, a redshank startled by our passing shrieks its 'Tu–u, tu, tu' in unconscious salute. We pass the lifeboat station and, keeping to the buoyed channel between the sand flats, make for the bar which guards its entrance.

For the first time *Elizabeth*, which has seemed rock-steady all the way down the channel, lifts with the sea. It is a gentle introduction to the short seas of the Southern Bight. The wind is south-westerly three to four on the Beaufort scale and we are in the lee of the land. Jack sets a compass course and glances at his watch but for the first part of our passage the coast is in clear sight and landmarks help to give pinpoint precision to our navigation: the Decca Navigator and Global Position Fixing by satellite are far in the future. Our steady progress north and east, our wake a bubbling furrow, eventually takes us out of sight of land. We can imagine its presence beyond the coastal haze, but as an aid to navigation its power is spent.

For me and for my father, the next two hours of steaming apparently blind, only to sight the first dan buoy marking the pots on the starboard bow, was virtually miraculous. Only later was I to learn at first hand the secret of dead reckoning, matching course, speed and time to arrive at a position. Jack held his course by making his corrections so subtly that our own wake was straighter than if we had been on automatic pilot, and so familiar was he with *Elizabeth* that the engine revolution reading told him all he needed to know about our speed. It was still a miracle. Tidal streams are fierce and complex off East Anglia, and allowing for them requires elaborate unconscious computation.

Apart from a Yarmouth trawler well to our east and steaming towards the Outer Silver Pit or the Dogger, we are apparently alone on the sea. The whelk boats that left with us are making for their own fleets of pots, each marked by a distinctive dan. The engine note changes as Jack eases down. The dan is alongside and, one by one, the pots are winched aboard to be emptied on to the deck and rebaited. A fringe of netting inside the entrance of each pot, known as a crinnie, discourages the catch from leaving. Not all of the catch is of whelks. Some apparent whelks start to scramble about to reveal the hermit crab within a dead shell. Lively butter fish, *Pholis gunnellus* (L.), thin blennies barely six inches long and with spots all down the back, squirm through the sides of the pots, having elected to stay aboard during the journey to the surface. The small, brilliant mauve sea urchins, *Psammechinus miliaris* (Gmelin), are shaken out with the whelks but their distant relatives, the common starfish, *Asterias rubens* L., and the occasional sun star, *Solaster papposus* (L.), often have to be pulled out bodily, resisting to the last with their myriad suckers.

This was to be my first meeting with echinoderms, one of the strangest yet most successful groups to emerge from the great pre-Cambrian genesis. Unlike most of the back-boned animals to which they are very distantly related, most have retained their body armour but in a form which permits growth without the need to moult. Although many animals, including ourselves, make use of erectile tissue at certain times, echinoderms have a full-scale water vascular system which powers the rows of tube feet, shod with suckers, on which they wander the seabed. As larvae, echinoderms are bilaterally symmetrical, as befits an organism moving forwards in a world where light and gravity distinguish up from down. With metamorphosis, the larvae assume all of the five-fold radial symmetry of a buttercup. It is a logical arrangement for a creature that moves slowly over the bed of the sea, in that progress towards prey, or away from threats, can be initiated at once without ever having to turn round. Looked at from our selfish viewpoint, in which we equate being advanced with being intelligent, radial symmetry has the signal disadvantage that it does not lead to the process of cephalization by which central nervous tissue from several body segments tends to be elaborated to form a brain at the leading end of many bilaterally symmetrical organisms. Nevertheless, in the great game of growth, survival and reproduction, which is the currency of all life on earth, the echinoderms have played in the First Division for over 500 million years. They show no sign yet of being ready to enter the relegation zone.

The whelks are bagged up and the uninvited guests, which by now include both velvet, *Necora puber* (L.), and other species of swimming crab, are shovelled over the side. We must steam on to the next fleet and the next until, after hours of back-breaking work for Jack and the crew, it is time to turn for home. The navigational miracle is repeated in reverse. Other whelk boats appear astern and abeam and, by a lucky chance, the nearest is the unofficial flagship, the *William Edward*.

We have seen a lot of gannet activity to seaward of where we hauled our pots, the great birds folding their wings like variable-geometry jets in the last moments before diving into the sea. Perhaps they were a sign that mackerel were driving sprats and young herring to the surface. There was no time to try for mackerel then, but the tide has still half an hour to make before we can cross the bar without grounding and we see the *William Edward* ease right down and splash feathered lines to port and starboard. In they come, a bright mackerel pulsing on nearly every feathered hook. We try as well but at first we are outwith the main shoal and just catch penny numbers. Suddenly we are in among the fish and soon their thick, muscular bodies are bounding among our sea boots. They are gone just as suddenly. We have caught a couple of boxes of a wonderful fish which, incredible though it seems now, could then hardly be given away.

Over the bar and up the channel: I am a real sailor at last – or so I think, any feeling of queasiness I experienced as the pots were hauled conveniently put to the back of my mind. Whelk-fishing gives only the tiniest glimpse of the wealth of the sea, but I have seen enough to know that I want to see a great deal more of it.

Botany and zoology were not taught at my school until the sixth form. Reaching this exalted level was, therefore, a double benefit, giving me the opportunity to study subjects of real interest in depth and the satisfaction of being treated almost as an adult, in my case rather prematurely. The school had a strong classical tradition but it was also well-equipped for the study of modern languages and the physical sciences. Budding biologists were looked upon as a little strange, not least because of the obscure location of the biology laboratory and the disorganized but entirely well-meaning personality of its Anglo-German head. He had wide biological interests but if he had a bias, it was towards the physiological and medical aspects of the subject. As a result, and in the face of my real inclinations, I began to think that the only way to make a living as a biologist would be in some branch of medical science.

But what I had briefly seen of hospital life as a patient did not appeal in the least. As far as I could tell, the only organisms present, apart from those causing diseases, were ill or injured patients, the environment was hot and dry, the food was awful and the all-pervading smell of disinfectant followed one everywhere. Thanks to an uncle, I owned a copy of Malcolm Smith's splendid contribution to the Collins New Naturalist series, *British Amphibians and Reptiles*. I gained some comfort from the fact that Dr Smith was a physician and that, if he could make the time to study healthy animals in the wild while earning his living tending to the needs of unhealthy members of his own species in captivity, then so, at a pinch, could I.

There I left the matter for the moment, but I knew that my real desire was to pursue my hitherto amateur interest in fisheries as a full-time occupation.

More than once during my seagoing life I have been told that I was 'tarred with luck'. Just as I was resigning myself to an indoor life pounding the over-heated corridors of hospitals, my Anglo-German mentor was replaced by a remarkable individual. Simon Lambert had been brought up near Keswick. As a very young man, he had been commissioned into the Indian Army where he saw action on the North-West Frontier against dissident Afridi tribesmen armed with machine-guns rather than the jezails, those long, muzzle-loading muskets made to a traditional pattern, of Kipling's day. Unlike Kipling's young officer, 'Two thousand pounds of education, shot like a rabbit in a ride', he survived this last British campaign on the Frontier. With the coming of Indian independence and partition, Captain Lambert was absorbed into the British Army as a subaltern in the Border Regiment, whereupon he decided to abandon his military career and to enter the University of Oxford under a scheme designed to attract young officers whose education had been interrupted by the Second World War. Here Lambert fell under the spell of Sir Alister Hardy, Linacre Professor of Zoology.

The influence of Sir Alister Hardy runs like a golden thread through the history of British fishery science in the twentieth century, and many of the figures who later came to dominate the subject were either former students of Hardy's or had been taught by those who were. Hardy's great contribution to marine science was to pioneer the study of plankton and to relate its seasonal abundance and vertical movements to the ecology of fish which, like the herring, depended upon it. The basis for his knowledge was his invention of a special piece of equipment, the continuous plankton recorder (CPR), which could be towed behind fishing and merchant vessels to trap plankton on a reel of cloth and preserve it in formalin for later analysis. This device and its predecessors were not Hardy's only research tools, or even the most important. His others were great enthusiasm for his subject and a wonderful capacity for getting on with the fishermen whose practical knowledge he greatly respected. To them, with his Oundle and Oxford education, he would have seemed like an exotic visitor from another world but, after he had been to sea with them, they quickly learned to return the respect he had for them. They admired his capacity for the hardest physical work and his ability to carry on under the worst of sea conditions. If he had much to learn from them about where fish were found and when, they were held spellbound by the explanations he was able to give on the basis of scientific studies in which they themselves had taken part.

Hardy's empathy with working men, which had nothing whatever to do with party politics, had its earliest expression in the relationship he formed with the fifty Northumbrian ex-miners he commanded in 1914. The survivors held reunions and exchanged Christmas cards with him until the end of his life. Hardy had a similar rapport with the ship's company of the Fishery Research Vessel (FRV) *George Bligh*, the Lowestoft-based ship aboard which he did much of his pioneering work. He kept a framed photograph of the ship and its people on his chimney-piece. It was there when he died and is now proudly displayed at the Lowestoft laboratory of what used to be the Ministry of Agriculture, Fisheries and Food but now has some complicated 'New' Labour name. Perhaps the most extraordinary expression of Hardy's relaxed relationship with what used to be called 'the working class' was his habit, while Professor of Zoology in the then new University College at Hull, of paying fish porters to spar with him at lunch-time.

We learned much about Hardy's original and attractive character from Simon Lambert, who had himself assisted with the analysis of CPR records and had taken an active part in fishery research programmes in waters as far apart as Zanzibar, the North Sea and the North-West Atlantic. That he had also worked as Educational Director of the Zoological Society of London at the time that Desmond Morris was Curator of Mammals, added to his credentials as a teacher of zoology *par excellence*. Within weeks, I had abandoned all thoughts of medical science and set course for a career in fisheries.

Hardy's tenure of the Oxford Chair marked the final flowering of what came to be called classical zoology, the study of the relationships between living and fossil animal groups as a discipline in itself and as a guide to the pathways taken by the processes of evolution. David Attenborough's *Life on Earth*, rightly described as the finest general natural history ever written, is a recent example of the value of the classical approach to the understanding of nature. The same approach was the basis of Lambert's teaching, which went way beyond the wooden and unimaginative limits set by the examination board.

With Lambert's arrival at the school came the commissioning of a superbly equipped biology laboratory, the design of which had been the responsibility of his predecessor. It included a vivarium, a room set aside as a potential zoo in miniature. The best we had

*A spectacled Cayman, the terror of the
school vivarium's frog population*

managed in the old laboratory had been a selection of aquaria
displaying lampreys, fishes and crayfish from the Chess. In the new
one it was possible to add a running-water aquarium for trout, and
vivaria for locally caught frogs, toads, common lizards, slow-worms,
grass snakes and even a spectacled cayman, a species of alligator,
from South America. The local mammals included short-tailed field
voles, wood and house mice and three specimens of the European
edible dormouse, part of a local feral population which had been
released over eighty years before by the first Lord Rothschild. Apart
from a pair of East African cane rats, the most exotic of our inmates
were two palm squirrels, a species of chipmunk, and two mon-
gooses brought back on separate occasions from India by a younger
fellow pupil.

Faced with such an embarrassment of biological riches, it is no
wonder that I put all of my academic energies into natural history
and only the necessary minimum into the other subjects required
for university entrance. I had especially little time for Latin, scoring
25 per cent at my first attempt and 20 per cent at the second. That
the Scottish Universities Entrance Board (a long-extinct organization from which one was required to obtain an oddly named
document called a Certificate of Attestation of Fitness) took pity

on one whose formal education was so narrowly based was a great surprise to my teachers – and to me.

─

A TASTE OF SCOTLAND

─

Among the older books in the family's sporting library, none was more eagerly read by me than Lord Walsingham and Sir Ralph Payne-Gallwey's two volumes on shooting in the *Badminton Library of Sports and Pastimes*. Of the two, the volume entitled *Moor and Marsh* was very much the favourite. Here were marvellous accounts of grouse and blackcock shooting, of 'fowling on the foreshore and of deer stalking among the crags of highland Scotland. It was quite clear that the opportunities for wild sport north of the border were in a different league from the homely fields and hedgerows of my native Buckinghamshire. Another early hint of the part that Scotland was to play in my life came from the family's friendship with the Hunter family of Pitchcott whose farm lay on a knoll overlooking the great sweep of the Vale of Aylesbury.

The Hunters had come to the Vale from near Mauchline in Ayrshire on the advice of Mr Hunter senior's doctor, who believed that the southern climate would be kinder to his patient's lungs which had been damaged by poison gas during the First World War. All of the family spoke the broadest Ayrshire, reared fiercely horned Ayrshire cattle and 'took the ploo roond the fairm' as if their part of the Vale had somehow been transferred to the west lowlands of Scotland. Fine ambassadors for their native land, they were hospitable to a fault and always keen to hear about our expeditions after the wild pheasants, duck and woodpigeons attracted to their broad acres.

On one day I shall never forget, my father and I were walking across a field of clover down on the low ground at Pitchcott when, to my alarm, an enormous cock pheasant rose at my feet with a great whirr of wings. It took me a few seconds to recover from the shock. Deciding what to do next was not straightforward. My father loaded his own cartridges and was a little jealous of their wanton expenditure by me. My first instinct was to pretend not to see the pheasant at all, but a glance to my left showed only too clearly that my father could see everything. So I carelessly threw up the little single-barrelled .410 and pulled the trigger. To my absolute astonishment, the cock pheasant fell stone-dead into the clover where it was pounced on and gently retrieved by Prince, our soft-mouthed but otherwise very wild liver-and-white springer spaniel. No one was prouder than my father.

One evening found us waiting by a little reed-fringed pond in the same field. A wild duck (only the drake is correctly termed a mallard) came in to be saluted by my brother with a 9mm garden gun, by me with the .410 and by my father's Thomas Bland twelve-bore. Down came the duck and, when the time came to pluck it, pellets from all three guns were generously reported by Papa. On another occasion, we were walking home in the gloaming across a great area of wheat stubble. A largeish, moth-like bird flittered past and, in a spirit of zoological inquiry that owed regrettably little to the Protection of Wild Birds Act of 1954, I shot it with the side-lever twelve-bore game gun by Stephen Grant to which I had by then graduated. I had bagged my first woodcock and it was to be many years before I bagged another.

It was, I suppose, inevitable that I would spend most of my adult life north of the border. There was the heady cocktail of wild sport promised by Walsingham and Payne-Gallwey, and there was also, of course, the wonderful example of Scottish hospitality set by the Hunter family. When, at the end of my schooldays, the time came to choose a university, I was delighted to hear that the oldest-established academic marine laboratory in Great Britain lay north of the border at St Andrews in Fife. The university of which it was a part had been founded in 1411, but the six papal bulls which granted its right to confer degrees were not issued by the relevant Pope until 28 August 1413. At that time, Christendom had a choice of three popes, Gregory XII and John XXII in Italy and Benedict XIII, who lived in exile in Peñiscola in Spain. Although he was regarded by some as the Antipope, it was to Benedict (Peter de Luna) that the Kings of Scots owed their allegiance and it was from the court of this pontiff that the bulls were issued. They finally arrived in St Andrews, amid great rejoicing, on 3 February 1414. My own arrival there, exactly 550 years after its foundation, was noticed by no one apart from the old lady whose case I carried over the station foot-bridge and who offered me my first tip – sixpence.

Some weeks prior to the start of term, the university lodgings officer had put me in touch with the owner of student accommodation in the town of St Andrews. It was clear from the businesslike tone of the landlady's letter that conditions in her establishment were somewhat regimented. That, the reference to 'plain tea' and the urban location put me off and I asked the lodgings officer whether she had a farmer or fisherman on her books. She had not but what she did have was the director of the Gatty Marine Laboratory whose wife was on the lookout for a student lodger to help keep an eye on her increasing family. Their house was at Boarhills and five miles along the coast from St Andrews, so I headed there aboard a country bus. It was blue and cream and bore the proud name of W. Alexander & Sons. My fellow passengers were jolly, rosy-cheeked housewives on their way home from shopping in St Andrews. To my gentle amusement, a fair crop of 'ye kens' and 'the noos' peppered their excited chatter but, unlike the Hunters of Ayrshire and Pitchcott, their Fife sentences always seemed to end at the upper end of the musical scale.

As the little bus laboured up the steep brae that takes the coast road out of St Andrews, I could see the fresh-ploughed soil shining in the crisp, long light of early October. Small fields enclosed by stone dykes led down to the cliffs overlooking St Andrews Bay, a few wind-blasted oaks and hawthorns taking the place of the great beech hangers of my native Buckinghamshire. A small grey stone kirk stood sentinel between the road and the sea, its simplicity broken only by an open belfry with its single bell. Beyond it, a squabbling flurry of black-headed gulls marked the path of a pair of Clydesdale horses cutting an arrow-straight furrow for their blue-overalled ploughman. The bus drew to a halt near a lethally sharp bend and Boarhills was announced. I reached for a brown pre-war suitcase and my father's old RAF kitbag from the luggage rack and stepped down on to the road leading through the village. Boarhills was then little more than four farm steadings and their cottar houses (tied cottages), a little post office incorporating a general store and a school for infant and primary children.

The road into the village led downhill until it turned into a steep uphill brae. At the top on the left lay the double cottage which was my destination. A shy knock and I was facing a tall young man in a faded tweed coat and Fair Isle jersey. To me he looked far too young to be a director of anything, let alone of a marine laboratory with a well-merited place in the history of British marine science and fisheries. When, in the well-modulated tones of Marlborough College and St John's College, Cambridge, Adrian Horridge introduced himself, I realized with relief that I could now put the kit-bag down.

MUD AND FEATHERS

At Boarhills, I quickly became used to grace before meals but I never fully got used to the sudden thrust of a dinner fork into my leg, usually delivered while heads were bowed by the young son of the house, to whom a paying guest was less than welcome. I was soon to learn that Adrian Horridge shared my interest in wildfowl-ing and rough shooting. He possessed a hammer gun made in Belgium before the First World War. It was chambered for the semi-magnum two-and-three-quarter-inch cartridge and was ideal for expeditions along the shore and, with a little discretion, certain areas inland. I needed no further encouragement.

I explained the attractions of Boarhills in a letter home. The indulgent response within a week was the despatch by my father of a twelve-bore shotgun by the American maker, Harrington & Richardson. It was a robust, machine-made hammer gun with a single, thirty-two-inch barrel chambered for the three-inch magnum cartridge. The gun was accompanied by a fair quantity of my father's home-loaded cartridges, including the awesome three-inch loads. Nowadays, of course, given the hysteria associated with anything to do with guns, the sending of such a package would trigger the arrival of an armed response unit at the home of both the sender and the recipient. Then, gun licences were freely available at the post office for ten shillings (50p) and you did not even need one of those, provided the gun was retained in one's house and 'the curtilage thereof'. Having shown me the local foreshore, Adrian was not at all possessive about my going there alone to try my luck. The day after the gun arrived, I hurried away from a practical class, changed into my old clothes and made my way down to the shore, the Harrington & Richardson over my shoulder and some magnum cartridges in my left-hand pocket.

It is still quite light as I walk along the track beside the neeps (swedes) and climb the stile on to the grassy slopes leading down to the sea. The light wind is offshore and, for a time, the smell of the neeps follows me downhill. A few black store cattle show a passing interest but, having established that I have no neeps to offer them, resume their munching of the seaside grass. There had been quite a storm earlier in the autumn and the strand line is marked by a thick, entangled mass of seaweed. It has started to rot and turn white and the smell of its decay soon takes over from the neeps. A 'scutty doo' (rock pigeon) clatters out of a cleft in a great local landmark known as the Buddo Rock, but I decide not to load up until I am in position among the rocky scars below Stony Wynd Farm. The tide is well out but still falling and I settle down in front of a large slab of sandstone overlooking a long rock pool. It is a lovely evening but cool with a dusting of snow on the distant peaks of the Grampians way beyond the other side of the bay. A few eider duck busy themselves offshore and, further out still, the primeval form of a cormorant skims the low wave-tops. Time to load up: I check that the bore is clear, insert a cartridge and gently close the breech. I pull down my face mask, put on my shooting mitts and await developments.

I hear a curlew and a redshank – both legal quarry then and excellent eating – but nothing comes within range. I am thinking of giving up when I suddenly become aware that a 'fowl of magnificent proportions has settled on the edge of the rock pool. I ease back the hammer, bring up the Harrington & Richardson and pull the trigger. A gout of orange flame stabs from the muzzle and one and a half ounces of number-five shot start their deadly journey. The back of my shoulder is first to strike the rock against which I have inadvertently been leaning and it is followed, a fraction of a second later, by my head. It takes a moment or two for my head to clear and the pain in my pinched shoulder shows no sign of abating.

Surely, though, such sacrifice will be rewarded? I slither forward on the wet rocks looking for the magnificent 'fowl in the grey dark. I find it sooner than I think, not much more than fifteen yards away. It is a redshank little bigger than a snipe. By a lucky chance, my aim was too high so that my quarry was struck only by a couple of pellets on the edge of the pattern. I slip it into my pocket and make my slow way home with ringing ears and an aching shoulder. A light bag is commonly the fate even of 'fowlers who can judge distance accurately and hold their guns straight. When I get back, I make a careful drawing of the redshank to enclose with my letter home, clean the gun and turn in. I pluck and draw the bird the following evening and add it to a mixed bag of other waterfowl shot by

Adrian, from which Audrey Horridge makes an excellent casserole.

The family who lived next door were called Kulik and their head, Stanislaus, was a young former member of General Anders's Polish army. Stanislaus – anglicized to Stanley since he married a local girl from St Monance and settled in Scotland – had been a paratrooper, a veteran of both Monte Cassino and Arnhem. His country had been badly let down by the Allies he and his comrades had fought with so gallantly and his view of Scotland was a little ambivalent. On the one hand, he was glad not to be subject to the dreary restrictions of life in a communist satellite, and on the other, he missed the hot, dry summers of his native land. As he put it, 'They call it bonny Scotland, but where is the bonny?'

It had been Stanley who had introduced Adrian to the local sporting opportunities – opportunities that some members of the local farming community might have described as poaching. One such activity, which was definitely legal, was flight shooting on the muddy estuary of the River Eden which enters St Andrews Bay just south of the Tay. Stanley had shown Adrian how it was possible to walk out into the middle of the mud banks by following a small creek that ran across the estuary about halfway down. The bed of the creek was navigable because the fine silt which rendered much of the estuary so treacherous was there washed away by the flow.

I had been taken several times to the estuary by Adrian in his black upright Ford Anglia, an excellent if unheated vehicle which boasted only three forward speeds but, because of its large wheels, was quite sure-footed off the road. My own contributions to the proceedings had been marked by many a loud bang but rather little to show for it. On our final trip to the estuary, Adrian very kindly suggested that, after the main flight was over, I should stay behind to engage any widgeon and waders foolish enough to fly near me as the tide rose and pushed the 'fowl off the far sands. Adrian then motored home to change and give a morning lecture while I remained on the *qui vive*, single-barrelled magnum in hand. For a time, all went well and I began to make a small bag. Eventually, when even the tide flight was over, I decided that it would be interesting to explore the estuary and, leaving the relative security of the creek, set out across the mud.

I think that I had plowtered on for about twenty yards only when first my right foot and then my left became inextricably stuck. I tried to pull one foot out and had just succeeded when I realized that the other, upon which I had had to put extra pressure, was absolutely immovable. Fortunately, the old gun was already unloaded and in its canvas cover. Laying it down on the mud, I slowly worked myself free by lying flat and making swimming movements. By the time I got back ashore, I was covered in black mud from head to foot and exuding that sulphurous smell known only to wildfowlers and other longshoremen. I made my way to the road and stopped the next Alexander's bus. It was crowded with standing passengers but, despite what must have been my desperate appearance, the driver allowed me to board. After St Andrews had been reached and seats became available, I nevertheless remained standing as a courtesy to fellow travellers. It was after this incident that Horridge's poor wife realized that not only was I no help whatsoever with the children or in the house, but also I encouraged her husband in activities she had hoped he had grown out of and spread mud and feathers wherever I went. Shortly afterwards, she arranged lodgings for me in St Andrews.

Nowadays, the old town of St Andrews, with its lovely buildings russet-streaked from the iron deposits in the local sandstone, is surrounded by a wide and ugly hinterland of wooden-framed monstrosities. Water that previously would have poured down field drains to feed the Kinnessburn and winnow the silt out of its gravel, is now redirected into the sewage system to re-emerge in the sea. The result has been such a reduction in the flow of the burn that the gravel where trout used to spawn is cemented with mud and the fish population choked off at source. How different it was during the summer evenings of 1962, when every pool bore the rings of rising fish. I had bought one of the first fibreglass trout rods, the Glaskona, made by Hardy's of Alnwick, and, after several weeks of frightening fish rising in water which was completely static, finally succeeded in landing a three-ounce sprat from the edge of a riffle. Such was the vigour with which I raised the rod tip that the poor trout had to be retrieved from the rank vegetation behind me. I am ashamed to say that it ended up fried in bacon fat back at my lodgings.

In the summer vacation of 1962, I was offered a six-week secondment at the Freshwater Fisheries Laboratory − formerly the Brown Trout Laboratory − at Pitlochry in Perthshire. My attachment was to be to a programme which sought to explore the relationship between the Atlantic salmon, *Salmo salar* L., and the fisheries which its populations sustained. The working day at the laboratory began promptly at half past eight, rather an early hour for the average undergraduate, and it was then that I was shown into the office of a Mr Kenneth Pyefinch, who bore the rather formal title of Officer-in-Charge.

Mr Pyefinch was not easy to make out, sitting as he was behind a large desk covered with what appeared to be the laboratory's entire filing system and puffing strongly on a briar pipe. He was by training and inclination a marine biologist whose early work had been on anti-fouling paints. It was by no means certain how strong his fisheries interests really were but none doubted his keen interest in railways and the fine detail of their timetables. His skills as a writer were beyond dispute, however, and his monograph, *Trout in Scotland*, based on the first ten years' work at Pitlochry, is still highly regarded.

Our chat was shortened by our mutual shyness but at the end of it I ventured to ask whether it might be possible for me to pitch my small tent in the laboratory's grounds. Clouds of tobacco smoke belched from behind the heap of files and, after a pause, I received a negative reply, to the effect that to do so would 'set a precedent' – in the eyes of the Civil Service of those days a crime second only to 'embarrassing a Minister'. I did not press the matter and pitched my tent at a camp site nearby. The highlight of my life there was the invasion of the tent by long-tailed field mice driven in by a fort-night of wet weather. They came to life at night, scuttling out from under the groundsheet and chasing each other over my prone body. Merry sprites, temporarily safe from the tawny owls that ruled the night sky and the weasels that sought them and their pink-skinned babies in their shallow runs, few would live longer than eighteen months. But for the genes, whose innocent vehicles they were, the brevity of their lives was so well compensated by high fecundity that there is more field-mouse DNA scurrying among the grass stems of Great Britain than DNA of any other small mammal.

As I had been given the mindless but important task of making plastic impressions of salmon scales using a tool like a miniature kitchen mangle, life at the laboratory was less fun than I had hoped. Every so often, relief would come in the form of a trip to the east coast with Willie Shearer. Willie was a child of Caithness in northern Scotland and his imagination had been caught as much by the human interest of the net fisheries for salmon as by the biology of the fish itself. It was his dream, which he was on the threshold of putting into practice, to make use of the unique opportunities presented by sampling the net catches to gain an understanding of the structure of salmon populations and the effects of fishing upon them.

We travelled to the east coast by Land-Rover. I chose to sit in the back to lessen the effect of the high-speed impact with a combine harvester which, in those pre-seatbelt days, my vivid imagination had put behind every blind corner. The east coast was the centre of the net fishery. In the days before the large-scale rearing of salmon, it was a substantial industry which, thanks to the interest and co-operation of the proprietors, provided excellent opportunities for Willie to put his programme in hand. In the final stages of their return migration from their sea feeding grounds in the North Atlantic, salmon seek the estuaries of their home rivers by swimming close in to the coast. Net fishermen exploit this behaviour by deploying curtains of netting, known as leaders, at right angles to the shore. There are several types of these fixed nets or engines, as they are called in the stained-glass language of this picturesque fishery, the choice of design depending on local conditions. They all work in a similar way. Fish encountering the leader seek to avoid it by swimming alongside it to apparent safety offshore, only to encounter simple netting devices which divert the salmon into a terminal trap or fish court.

Part of a haul of post-smolt Atlantic salmon taken in the Faeroe–Shetland Channel in late spring 1996

Salmon and sea trout from the various fixed engines and taken by net and coble – a form of beach seining – from within the estuary and lower river were brought to a great fish-house in California Street, Montrose. To those who have never known anything but the relative wild-salmon scarcity of the last quarter of a century, the scene in that fish-house would have seemed breathtaking. Rank upon rank of fish from sea trout of two to three pounds to great summer salmon, some weighing over thirty or even forty pounds, were speedily iced and boxed. Such was the abundance that some were even set aside and quick-frozen for later sale during the close season. Everywhere the curious, biscuity smell of salmon skin invaded the nostrils, a scent almost certainly important to the fish in life and one which clung to one's hands and clothes for hours. What made the scene in the fish-house all the more impressive was that the populations of salmon, of which these catches were but a sample, were also supporting a new fishery by drift net offshore.

These new fisheries were made possible by the use of monofilament nylon in the construction of drift nets. Unlike the traditional gear made of braided natural fibre, the new drift nets were difficult for the fish to see and quickly demonstrated their efficiency. To experienced fishery biologists, sudden increases in fishing efficiency ring loud alarm bells. The abundance in the fish-house was an indication of a relatively intensive fishery, at least locally. However, the good representation of large older fish in the catch, including those which had spawned more than once, suggested that the salmon populations supporting the fishery were still in good heart. But how would the stock stand up to the additional exploitation by the new fishing gear? We knew that the effects of this increase in fishing pressure might take one or more salmon generations (five to ten years) to make itself felt. It was not the only new problem. Far away, off the coast of West Greenland, other drift and gill nets made of the new material were also taking salmon, some of which were undoubtedly of Scottish origin. We had every reason to be concerned, but in the meantime we had a job to do, alternating sampling at the fish-house with trips to sea with the local drift netters.

The vessels using the new nets had depended for many years on fishing the seabed for fish of the cod family (gadoids in the language of the biologist). Of these fishes, none was more important than the haddock, *Melanogrammus aeglefinus* (L.), mainstay of Scotland's other national dish, the fish supper. The snag in the fishing season of 1962 was that haddock stocks were at a low ebb because of several poor brood years. Naturally wide variation – even up to a factor of 100 – is normal in the haddock and is not, in itself, a cause for concern. To the fishermen, though, this period of scarcity was causing real hardship for which the rising price of the fish could not fully compensate. The advent of a new fishery based on salmon provided welcome relief for them but consternation and outright hostility from the traditional salmon men. As biologists attached to the Government, we were obliged to stay clear of such squabbles, however much we valued the personal friendships we enjoyed with members of both fishing communities. Our rightful concern was instead the capacity of the stock to withstand losses from all causes including natural ones, and this was a question that we did not then have the wherewithal to answer.

The little harbour at Gourdon was the last stronghold of sma' line (small-line) fishing in Scotland. Because the fish had not been churned around in the cod end of a trawl or seine net, they were of the highest quality and fetched a premium price. The snag was that line-fishing was highly labour-intensive, even if the fishermen had access to what they regarded as free labour – their wives – to bait the lines with mussels. The small decked boats used by the line fishermen had low-horsepower engines so trawl-fishing, other than on the smallest of scales, was not really a viable alternative to working the lines. Drift-netting for salmon was another matter altogether and there is no doubt that many of the line men saw it as a gift from on high.

We arrive at the little harbour after the usual white-knuckle ride in the Land-Rover. Although it is a clear, sunny day inland, a 'haar' or sea mist invests the coast and is refreshingly cool after the heat and noise of the vehicle. We are here to observe the fishery at first hand and especially to see whether salmon taken alive from the drift net would be suitable subjects for tagging. With the impatience of youth, I am keen to meet our crew but, in the measured words of my companion, who realizes that the fishermen are doing us a favour rather than the other way round, 'We must await their pleasure.' We do not have long to wait. A wiry little figure wearing a reddish smock over his blue jersey makes himself known and invites us aboard his trim little liner, its wooden hull protected from the red-grey stone of the harbour wall by car-tyre fenders.

Fishermen, especially those from the closed traditional communities of north-east Scotland, can sometimes be somewhat taciturn when meeting strangers for the first time, but not so our new friend. His eyes literally sparkle in the light of the wheel-house as he tells us about a wonderful 'shottie' they had the previous week when they took over a hundred grilse (salmon that have returned after just over a year at sea) from the nets. Then the little Gardner diesel rumbles into life and we nose out of the harbour into the haar, the late evening silent but for the sound of distant foghorns. The nearest nets, with their be-flagged dan buoys at each end, are about five miles from the harbour. There is almost no wind but there is a long swell from poorer weather farther to the north, 'the dog before its master'. We steam along the net and see the struggling form of a single fish: it is time to lie off during the short hours of darkness that mark high summer in Scotland and wait to see whether the previous week's bonanza will be repeated.

Lying hove-to in a long swell affects different people in different ways, the fishermen not at all. I start to feel sleepy and my colleague vomits quietly downwind. There is a lot more light now and the

Gardner resumes its rhythmic throbbing. We are back alongside the net and, after establishing that we have caught but a handful of grilse, the skipper decides to haul anyway. All of them are alive bar one, perhaps the first one we saw, but they are in no state to be tagged. The pitiless nylon twine has held them partway down their struggling bodies, displacing scales and bruising the flesh within. They are not the only fish to come over the side. A school of mackerel, *Scomber scombrus* L., has slammed into the net at speed, embedding the nylon twine into the angle of their jaws. The only other fish, a lemon sole, *Microstomus kitt* (Walbaum), has somehow entangled itself right in the top of the net, a surprising end for a bottom-living flatfish. If fish had been the only catch, we could have returned to shore, if not content, then at least with our consciences clear. Sadly, a group of guillemots, perhaps chasing the same sile (sprat and herring fry) as the mackerel, are also in the net. They are trapped by their wings which are cut by the cruel nylon. All are dead.

My next trip to the coast was to be in the lively company of Kenneth Balmain. He was a child of North Berwick, where his late father had been a photographer in direct line of succession from the great Scottish pioneers of that profession. There was much of the artist about Kenneth as well. His grey hair was longer than was normal for those days and he wore a neatly trimmed goatee beard. He was an expert at tying his own flies. He loved all forms of fishing but, if he had a favourite, it was taking the wild brown trout of the River Tummel on a single dry fly fished upstream. If Willie Shearer was to be the Freshwater Fisheries Laboratory's ambassador to the traditional netsmen, Kenneth represented the cause of science to the angling community. They respected him for his angling skills and knowledge of how to improve fishing opportunities by putting science to work. They liked him because he was always prepared to listen and to enjoy a dram or cigarette with them late into the evening.

We set off, not in a Land-Rover, but in the greater comfort of a van, the tinny body of which vibrates at a higher frequency than the heavier vehicle's. That and the slower pace make normal conversation possible and, even from my seat in the front, I no longer anticipate tearing metal and oblivion round every blind corner. We are on our way to Arbroath in Angus to join another old wooden vessel. It is larger than the Gourdon sma' liner. Kenneth thinks that it was probably built for the herring fishery just after the First World War, although it might be older. It is equipped for light trawling but for the last few weeks it has been making its living 'at the red fish'. ('Red fish' is a euphemism for salmon in the traditional and highly superstitious world of the fishing community, which believes that the inadvertent mention of certain words at sea will bring bad luck.)

We are invited down to the forepeak where a coke-burning stove has turned a warm summer evening into a sort of fishy Black Hole of Calcutta. Once again, we hear about the scarcity of haddock and the boon that the new fishery has been. It is some time before we are due to sail and, to escape the fug for a few precious minutes, I clamber back on deck. It is a lucky moment to look out to sea because there on the horizon is a large, square-rigged ship in full sail. There is no large British square-rigger currently in commission and, being ship-rigged – that is to say, a three-masted vessel with yardarms on all three masts – it cannot be the German navy's *Gorch*

Foch. It is probably a Russian training ship, and a great deal newer than it looks.

I am back below-decks when the engine starts and, just as someone hands me a large plate of baked beans and fat bacon, am nearly thrown off my feet as the whole world plunges and corkscrews in a short head sea. Kenneth handles his plate with the easy nonchalance of the experienced small-boat sailor. I do my best to look grateful when a mug of cocoa is pressed into my right hand and try to ignore the olfactory combination of fish, diesel fumes and cigarette smoke that pervades our little world. By a combination of restricted breathing and sheer willpower, I clear the last fatty piece of rind from my plate and take the cocoa up on deck. Thank God to be out of there and holding on to the rail on the lee side of the wheel-house. The wind is getting up and we are steaming at our full speed of nine knots to haul the nets before the weather gets any worse. But it is no longer comfortable on deck and I return below to find Kenneth smoking a cigarette and in lively conversation with the deck-hands. I have yet to acquire the skill of making unconscious allowance for the motion of a fishing boat in a seaway and climb gratefully into a bunk. To my surprise and relief, my queasiness recedes and I nod off.

I awake to the sound of thumping noises on the other side of the bulkhead, grab my oilskins and return to the deck. The net is coming aboard and the thumps are the sound of salmon being thrown down into the fish-room where they are being packed in ice. The mackerel are tossed into fish boxes on deck and the bodies of the drowned birds are roughly pulled from the meshes and put over the side. What a waste – in Iceland they would be skinned and sold as delicacies. The youngest deck-hand is clearly as shocked as we are by all the dead guillemots, but the worst shock is yet to come. A school of 'puffies' (porpoises) has run into the net. Two have stuck fast and drowned, confirming in Kenneth's mind and mine that, whatever the rights and wrongs of catching salmon in this way, the hidden costs of this fishery are just too high.

Mercifully, the Secretary of State for Scotland was to approve a temporary prohibition on 15 September 1962, one which was made permanent the following year. By a lucky chance, haddock reproduction that year was exceptionally prolific. It was the start of the so-called gadoid outburst, a climatically driven phenomenon which, for a time, restored prosperity to the white fisheries and softened the effect of the ban on drift-netting for salmon. Incredibly, however, forty years later a remnant of this cruel fishery still continues with legal blessing off north-east England and, on a considerably larger scale, off the west coast of Ireland.

In that summer of 1962, the tortured imaginings that stayed with me were of the death struggles of the two porpoises which, only the day before, could well have been sporting joyously around the bluff bows of the fishing boats whose deadly new nets had been their nemesis. How lifelessly were their gleaming bodies slopping back and forth as each wave caught them; how bloodied their permanently smiling faces and how cruelly scored their flawless skins. Were such melancholy thoughts behind the commendably swift decision to ban the fishery? Of course not. The ban was prompted both by *realpolitik* – the outrage of the established long-shore netsmen had to be placated – and the concern that, even at a time of high abundance, adding yet another hurdle for the return-ing salmon stocks to overcome could endanger their capacity to reproduce themselves in the river. Did biological science have any contribution to make? Only one, and it was forcibly made. There was incontrovertible evidence that most salmon return to their natal rivers at spawning time. The best way to manage their populations is, therefore, to match the numbers of spawners to the capacity of each river to support their progeny. Intercepting the mixed populations of salmon in the open sea makes such fine tuning impossible. It was an argument which Scottish officials and politicians were glad to use.

Why was the same swift action not taken in north-east England? It was because the *realpolitik*, but not the science, had a different face. Drift-net fishing, using inefficient nets made from natural fibres that the fish could see, had a long history in Northumbria. In a region of high unemployment, the greatly invigorated fishery was welcomed by politicians anxious to hold on to their marginal seats. As to the science, the same points were made as in Scotland. (More damningly, the results of tagging experiments clearly demonstrated that most of the catch was of salmon intercepted on their way north to the rivers of eastern Scotland.) But I had learned the hard lesson that, in Winston Churchill's famous wartime words, 'Scientists should be on tap and not on top'. Much later, I was to draw the cynical conclusion that, to many politicians, scientific evidence is merely something you use either as an excuse to do nothing or as a justification for doing something unpopular.

STRAIGHTER POWDER

One of the secondary reasons for my taking the job at the Freshwater Fisheries Laboratory was to finance an expedition to the west coast of Scotland with my brother Peter. The sense of loss I had felt when my father sold his magnificent eight-bore many years before had never left me, but now we were to visit Salen in Argyll and view an eight-bore, on sale for what was then the princely sum of £25. My pay at the laboratory had been £6 1s a week, so financing the whole operation required certain economies. These included making full use of the food items such as stale bread and rolls discarded by holiday-makers sharing the camp site with me, and guarding against scurvy by eating young bracken shoots, of whose carcinogenic properties I was then quite unaware. In the short term, my brother was concerned, not to say shocked, that I had gone native in this way. We set off for the west coast the day after his arrival and found it full of midges and the most charming of people. Viewing the eight-bore involved a trip across Loch Sunart in a tarred, clinker-built boat rowed by fishermen singing quietly in Gaelic. As a gesture to us, they ventured to say that 'Indeed, the midges are bad today'.

We left the fishermen to clamber up a stony beach and made our way through a bog strewn with northern marsh orchids and cotton-grass to the cottage where the mighty eight-bore was said to reside. A shy knock on the door and we were ushered in to the front room. Given our disreputable appearance, this was a real tribute to the tradition of unquestioning hospitality so characteristic of the West Highlands both then and now. I am not sure what we expected. I had, of course, handled my father's Purdey eight-bore, and I had seen a photograph of one other, made, I think, for the Maharajah of Jaipur. We sat expectantly, to await the advent of the tooled oak and leather case in which, or so we imagined, the Argyll eight-bore was normally kept. Our host re-entered. A pair of rather rusty thirty-six-inch barrels projected from under his arm and two great hammers peeped from the crook of his elbow. It was certainly a prodigious weapon, the leather pad roughly nailed to the stock testimony to its powers of recoil. I opened the gun by moving its sturdy Jones under-lever to the right. One look down the barrels was enough. Just forward of the chamber of each was a 'pit' – a jagged hollow in the bore caused by rust – one could have lain down in. The Holy Grail had eluded us and, with money fast running out, it was time to return home to Buckinghamshire.

Within a few short weeks, I would be back at King's Cross Station on board the Night Scotsman. By the time I was back in Fife, the 'fowling season would have started, but how would I pursue it? The old single-barrelled Harrington & Richardson was no longer really an option. The three-inch cartridges had all been expended, many in vain, and the extractor spring was now so weak that the only way to remove a cartridge, live or fired, was to drop a weight down from the muzzle – not an especially safe proceeding and, in any case, difficult to do when crouched on mud or among rocks.

I still had, however, the Stephen Grant game gun with which I had shot the woodcock some years before. It was, and is, a most wonderful-looking gun, handmade in the closing years of the nine-teenth century when British gun-making was the envy of the world. Like the cathedral builders of the Middle Ages, Stephen Grant and his craftsmen had had absolutely no training in formal design, yet they had achieved, in their sublime product, a wonderful marriage of form and function. Externally, there is not a straight edge anywhere, from the tapered barrels via the 'fences' (the demispheres that form the breech) with their 'hare's lugs' carved in unconscious recollection of the muzzle-loading era, to the snake-like reverse curve of the side-lever which opens the gun. Small wonder that the identical model is illustrated on page 90 of *Shooting in Field and Covert* in the *Badminton Library* as 'a well-shaped gun by Stephen Grant', or that in Adam's and Braden's *Lock, Stock and Barrel*, a tribute to British gun-making by two Texans, we read that 'God shoots with a Stephen Grant'.

Unlike the Harrington & Richardson, the Stephen Grant fitted me perfectly, shooting exactly where I looked. Like a lot of fine guns of that era, it had seen much service. The action wore its years lightly, but the barrels were another story. They had worn thin and corrosion near the muzzle had created a small hole. The price of a new pair of barrels would have been prohibitive but fortunately the renowned gun-making firm of Westley Richards had invented 'resleeving', a process in which the old barrels are sawn off near the breech and new tubes inserted. The gun was duly sent away and returned after a remarkably short time with a new lease of life. This was the gun I was to take back with me to St Andrews. I never exposed the Grant to the mud of the estuary, but it was to be my

'The gun God shoots with.'

companion of the open coast. A scutty doo exploding from a cleft in the rock was no longer totally safe from the right barrel, and the heavily choked left had a long reach over the sea.

Sometimes, as I waited among the rocks for the light to go and the curlew and mallard to take flight, my only company would be the local fleet of creel boats, direct successors of the line fishers of the Victorian era but fishing for crabs and lobsters. Their boats at that time were built to a lovely traditional design reminiscent in miniature of the sailing 'Fifies' of the nineteenth century. Like them, they were carvel-built of larch on oak, a system of building which leaves the surface of the hull smooth. It is also much more robust than the ancient clinker build in which the planks that form the hull overlap one another and tend to spring apart if struck from the side, a real risk when hauling creels close inshore. There was no shelter for the crew of these boats in their thick Fearnaught trousers and yellow oilskin smocks, and neither were the boats fully decked. The creels were hauled with the aid of a simple powered capstan mounted on the gunwale. So far as I could see from the shore, crabs were the main catch but, every so often, a blue-shelled lobster would hold up proceedings. Lobsters were much more valuable than crabs but were inclined to damage one another if their claws were not first banded with strong elastic. Soon, these strange animals and I were to get much better acquainted.

BETTER A LOBSTER THAN
NO HUSBAND

It is probably best not to inquire too closely into what precisely is meant by the old Gaelic proverb, '*Is fheàrr an giomach na 'bhi gun fear tighe*' ('Better a lobster than no husband'). Heaven forfend that it should have the same anatomical connotation as 'widow's comforter', the colloquial name for the pork sausage in parts of the English West Country. I fear, however, that it has.

My own initiation into an understanding of the life – and possible uses – of lobsters was to be less controversial. It was to take place at the Gatty Marine Laboratory at St Andrews, where, by a series of lucky accidents, I had succeeded in obtaining a research grant. I soon got over my initial disappointment that the Gatty was no longer a fisheries laboratory but merely a place where marine animals were used to study essentially physiological problems. Nevertheless, when the time came to choose a research topic, I was keen that it should have some relevance to the fishing world in which I hoped eventually to make a career.

My salvation was the appointment of a remarkable young lecturer, Mike Laverack, who, having tired of studying earthworms under rather dull circumstances in Grange-over-Sands in Lancashire, was keen to try something more stimulating. Adrian Horridge, my former landlord and the director of the Gatty, had suggested to him that rather little was known about the sense-organ systems of lobsters and crabs, animals apparently insulated from the outside world by a rigid armour. Mike's early work with these large crustaceans had shown that they possessed small, delicately articulated, branched setae or cuticular hairs which enabled them to detect the strength and direction of water movements so fine as to give them a sense of touch at a distance. This exploitation of what acoustics experts call the near-field effect is a characteristic that had only been fully described before in fish and amphibians. Perhaps, if we looked a little harder, we would uncover other fish-like properties in crustaceans. As far I was concerned, it was very definitely a case of 'better a lobster than no fish'.

Mike Laverack's father, a Lancashire Fusilier, had been severely wounded in the fiasco of the Gallipoli landings in 1915 and so was unable to take as active a part in his sons' lives as he would have liked. That handicap, and an upbringing in Croydon, denied Mike the first-hand familiarity with the natural world that had been the privilege of my young life. It was perhaps because of this that his early biological interests tended not to go beyond the experimental investigation of parts of animals in the laboratory. On the face of it, my own interests in the ecology of whole animals, especially of fish and shellfish, were the exact opposite. In fact, of course, the two approaches were merely opposite sides of the same coin. My little project was to take a further look at some of the other mechanisms that a hard-shelled animal like a lobster uses to make sense of the world around it. But the closest I had been to a live lobster was the hundred or so yards that separated a wildfowler among the rocks from a fishing boat offshore. I had a lot of reading to do and started by trying to learn a little more about the evolution of the Arthropoda – literally, jointed feet – that vast group of animals of which lobsters and crabs are advanced members.

•

Lobsters are extraordinary animals, close marine relatives of the freshwater crayfish of my childhood. Like them, they are very dependent on stones and rocks to provide shelter when they are moulting and areas of lee in which they can seek their prey without being swept away by currents. If our native crayfish rarely grow bigger than a man's hand, lobsters over two feet in length and over fifteen pounds in weight are by no means unknown. How big they get is as much a function of their survival as their growth-rate. For those lobsters lucky enough to live to be too large to enter the creels and pots of the fishermen, their subsequent growth depends on the availability of shelter.

The European lobster,
Homarus gammarus *(L.)*

As the lobster gets bigger, and its body projects more and more into the faster-moving water above the boundary layer at the bottom of the sea, so the risk of its being swept away increases. Furthermore, when they moult, lobsters renew their entire skeleton. In the hours after moulting, a lobster's body is as soft as a newly turned-out jelly, and its need for secure shelter is literally a matter of life and death. During this period, the lack of a large enough space under a stone or rocky crevice can bring the life of even the most formidable lobster to a premature end.

If this is a dangerous time for the animal, it is also a great nuisance to the fishery scientist because it leaves him with no hard structure like a scale, tooth or bone from which he might be able to determine its age. We therefore have no means of knowing the exact age of any one wild lobster or, indeed, of any other crustacean. However, like many other animals including ourselves, the cells of lobsters accumulate yellowish age pigments called lipofuscins which build up steadily as the years pass. Recently, my brother Peter took part in a study in which he and his collaborators attempted to estimate the age of lobsters from the level of lipofuscins in their cells. Their results surprised them. It seemed as though it could take as long as seven years for a lobster to reach the minimum size at which it could be landed, of just over three-quarters of a pound. They were also surprised to estimate the age of one two-foot Yorkshire giant at just over seventy years. Clearly, lobsters and their relatives are remarkable creatures. How had they come to be?

With the discovery of the biochemical basis for the genetic code has come the realization that evolution is about how, over time, selfish genes have developed increasingly elaborate means for perpetuating themselves. The earliest living things, of which we have certain knowledge from terrestrial rocks, are bacterium-like organisms in which the genetic material is suspended freely in a single outer membrane. Such organisms are known as prokaryotes. In the so-called eukaryotic cells of which all animals and plants are made, most of the genetic material is contained within a special structure, the nucleus, separated from the rest of the cell by its own membrane. Eukaryotes appear to have resulted from successful symbioses between two or more prokaryotic predecessors. Later, increasingly elaborate combinations of eukaryotic cells were sometimes found to be more successful at replicating the genes they were carrying than their single-celled ancestors. So the process continued until animals evolved which were not just combinations of highly differentiated groups of cells but combinations of such groups, called segments. In the more advanced types, a common gut linked the segments and the front end became increasingly modified to facilitate feeding and to enable information about the outside world to be received and processed. One such group became the true 'worms' or Annelida and it is from this group that the ancestors of all arthropods, including lobsters, are thought to have evolved.

Look inside a worm or certain primitive arthropods and it is clear that, over much of the animal's body, one segment is much like another. In each segment, sense organs, muscles and appendages are linked by nerves to a sort of junction box called a ganglion which is connected to its fellows in other segments by a ladder-like nerve cord. In advanced arthropods like lobsters, there is a great deal of specialization among the segments, some carrying antennae, eyes or mouth parts and others claws, walking legs and gills. Along with this local specialization goes a great deal of front-end compression to form a combined head and chest or cephalo-thorax. The most important specialization of all is that the thin, flexible cuticle inherited from worms is thickened and hardened to form a jointed armour. Just like the body of a modern car, this monocoque structure provides both rigidity and protection. What lobsters and crabs do, including how they locate and climb into fishermen's creels, is largely the story of what they can feel through their armour and how, in reacting to it, they move one armoured structure relative to another.

Our interest was ostensibly in the processes by which information about events in the world outside these crustaceans was conveyed inside to the complex of ganglia which form their central nervous system. I say 'ostensibly' because projections from the armour extend inside to form attachments for muscles analogous to our ligaments, and other bands of connective tissue span the joints. These bands are elastic and contain groups of large nerve cells, the responses from which provide the central nervous system with information about the tension in the joint and thereby the relative position of the hard structures (e.g., pieces of leg or body segments) on either side. Why does any of this matter to a lobster or crab fisherman? It matters because lobsters and crabs are not caught passively by a towed net. They walk into the creels themselves in response to the smell of the bait of dead fish inside. Detecting the smell requires a chemical sense; knowing which way to walk requires information about the direction of the current; and the action of walking itself requires information about the relative positions of leg and body joints.

One of the problems with being a large creepy-crawly is that there are lots of bits sticking out for predators to grab. But if a seal or a large fish grabs a lobster's leg or claw, it gets a surprise. Special muscles contract around a fracture zone at the base of the limb. The lobster escapes, having cut off the blood supply to the cast-off limb, and the predator is left with an *hors d'oeuvre* rather than a main course. Over the next few months, the lobster grows another limb and, with luck, lives to a ripe old age. We were able to exploit the lobster's trick in our work because all of the sense organs in which we were interested were present in isolated limbs. All we had to do was to ask a lobster for a leg, connect it up to our electrical equipment and prepare to publish the results in a leading scientific journal.

The starting point for the investigation was to be a group of large cells at the tip of the leg. They had been discovered, and described as chemical sense organs, by Wilhelm Luther in 1930. Large nerve cells produce correspondingly large electrical signals, so there was a good chance that not only would we be able to confirm the discovery directly, but also to learn more about the process of chemo-reception itself.

My knowledge of electronics then and now could be written on the back of a postage stamp but, over a period of weeks and with an enormous amount of help from the laboratory's technical staff, I somehow assembled enough kit, and in the right order, to both amplify and display any signals the cells were likely to produce. The structure to which the cells were connected was the horny tip of the leg. In some ways it resembles a mammalian claw and, like a claw, is slightly flexible. The ends of the nerve cells could be seen to enter canals within the claw, canals that were believed to hold the secret of chemo-reception.

Came the great day, extensive tests had shown that the electronic ironmongery was working perfectly and a fresh lobster leg lay in position. Carefully I dissected out a fine bundle of nerve fibres with the finest of tungsten needles and lifted them out of the saline on to the paired electrodes. I looked up to the green flickery lines on the oscilloscope screen (it was of the very latest type, expensively imported from the American West Coast) and centred the display. It looked good: could earning a Nobel Prize really be this easy?

I turned my attention back to the tip of the leg. I had even gone to the trouble of isolating the tip with balloon rubber so that, when I applied the chemical stimulus from a dropper, it would not contaminate the saline in which the nerves were bathed.

The common shore mussel, *Mytilus edulis* L., is highly attractive to lobsters and I had made up and filtered an extract of its tissues to apply to the tip of the leg. I could not look at the dropper and the screen of the oscilloscope at the same time. My technical mentor had thought of this and had added a small loudspeaker to the kit so that I could hear any electrical discharges from the nerve cells as well as see them. I positioned the fine pipette over the tip of the leg. The gentlest of squeezes and a crystal-clear droplet of pure mussel juice kissed the target. Dead silence: I cleaned the preparation with filtered sea water and tried again. The result was no different: perhaps, in my ignorance, I had failed to connect the equipment together properly or had got the amplifier on the wrong setting. I checked everything and found nothing out of place. It was time to have another go, only this time I contrived to keep my eye on the oscilloscope screen while a colleague applied the mussel juice. Not only did the microphone remain silent but the screen display remained as flat as a penny.

Week succeeded week, the apparatus was stripped and reassembled and I performed prodigies of dissection with my tungsten needles. Perhaps I should try a change of career? One thing was certain: I hated this indoor science and increasingly found myself looking out of my window at the fields of Braid's Farm next door. Sometimes lapwings would tumble above the stubble, to settle and be joined by little groups of golden plover. On this morning the field was bare but just then a single 'cushy doo' or woodpigeon, its eye no doubt caught by a last grain of spilled barley from the little combine, glided in and landed softly. It was about thirty yards away. Later in the day I had planned to indulge my need to get outdoors by going to Boarhills in search of rabbits. I had brought my gun into the laboratory and, picking it up from the corner of the room, I eased open the window. I slipped a cartridge into the left barrel, gently closing the action very correctly by raising the stock up to the barrels. The doo was still busy with the grain. I took a good aim, eased the safety-catch forward and pulled the trigger. The report of a twelve-bore is surprisingly loud to those unused to it. The report of a twelve-bore fired indoors is louder again. My door shot open and a white-faced Laverack rushed in, convinced that a major item of apparatus had exploded. His reaction to what had really occurred was less sympathetic than it might have been. I put this down to his urban upbringing and said no more about it. Needless to say, the doo was eaten that very evening, with some potatoes I had found dumped over a cliff by a farmer, and some green vegetables in the form of nettle tops.

My crustaceans, however, were still keeping their secrets. It was to be some time before the penny dropped that, if there were any chemically sensitive organs in the lowermost joint of lobsters' and crabs' legs, they were not at the tip. Confirmation came when, by accident, a drop from the pipette landed on one of the rows of setae a little further up the leg. There was a sudden, crackling buzz from the little loudspeaker and I looked up at the oscilloscope screen. The green line had been replaced by what looked like a bottle-brush as bundles of small nerve fibres repeatedly discharged. Steadily they died away, disappearing altogether when I washed the setae carefully with filtered sea water. I repeated the trial, this time with a camera attached to the screen. Had it been properly loaded, I dare say I would have had something to show the world. That had to wait until the following morning and the help of a colleague with fewer thumbs and hang-ups about modern technology. By coffee break, figure one of my thesis was drying under the window.

I could tell from the height of the spikes on the oscilloscope screen that the nerve cells from which I was recording, although very numerous, must also be very small. This conclusion begged the question: what were the massive cells in the tips doing if they were not sensitive to chemical stimuli? I wasted no further time bathing them in filtered mussel juice and tested the hypothesis that they were sensitive to small mechanical deformation of the horny tip. The result was immediate. 'Click, click', 'click, click' – great spikes appeared on the oscilloscope screen, spikes so large that they were off-scale and I had to turn down the gain control so that I could photograph them. After a whole winter of frustration, I had made two critical discoveries in two days. The task that lay ahead was to repeat and extend the work under controlled conditions. It was a job that would take two rewarding years, entirely uninterrupted by salvoes from the window.

THE SMILING DON

Apart from disabusing me of the naïve idea that the practice of science was easy, one of the other benefits of these months of distress and discontent was to hone my skills in the subdividing of nerve bundles. Now that I knew the real function of the large nerve cells in the tip of the lobster's leg, I was anxious to put my hard-won dexterity to use. The following morning, I lit the Bunsen burner as usual and waited for the mineral I used to sharpen my tungsten needles to melt in a white ceramic crucible rather like a little eggcup. The crucible had a rounded shape and tended to wobble slightly in the turbulent blue flame of the Bunsen. As I waited for the mineral to melt – I think it might have been caustic soda – I cast my eye out of the window. Heavy, drenching rain, of the sort the Irish call 'liquid sunshine', was forming large puddles in the field and I wondered whether they would remain long enough to attract the mallard that flighted out from the Kinnessburn every evening. When I looked at the crucible again, it was glowing red and its contents were bubbling fiercely. As I rushed to turn off the gas tap to the Bunsen, the sleeve of my laboratory coat caught the edge of the square wire-and-asbestos grid on which the crucible was balanced. Its molten contents splashed partly on to the bench and partly on to the thumb and index finger of my right hand. The pain was so excruciating that I was not even able to string any swear-words together.

It took some days for the burns to settle down, but I was so keen to learn more about the large nerve cells that I found myself back at my binocular microscope, subdividing the nerve bundles of the leg, long before my hand had healed. I found that, in the best-dissected preparations, one cell would discharge on pressure being applied and another smaller cell on its being released. It looked as though each canal in the tip of the leg contained two nerve cells. Later, I was to see them directly under the microscope and was able to confirm that the sensory cells were indeed paired and that one of each pair was larger than its fellow. Similar arrangements of sensory nerve cells could be seen embedded in the elastic tissue which spans each of the leg joints. Although this tissue is deep within each joint, it is nevertheless part of the animal's cuticular skeleton. This specialized tissue has a similar function to our own knee ligaments. The function of the pairs of sensory cells in the lobster's elastic tissue is to tell its central nervous system where each joint of the leg is located relative to its fellow.

In the receptors I was looking at, the elastic cuticle was on the outside of the animal and responded, not to the movement of the next, non-existent leg joint, but to movements of the tip of the leg relative to any hard objects with which the walking lobster made contact. It seemed to me that these receptors were doing a similar job on the outside of the animal to that done by the stretch receptors on the inside of the leg joints. Indeed, in more speculative moments, I wondered whether the internal stretch receptors had evolved back in pre-Cambrian times from external structures not so very different from the ones I was looking at. Perhaps one day the discovery of a new fossil, or even conceivably some archaic form still living in the obscurities of the deep ocean, would answer my question. However, of one thing I was certain: lobsters were much more sensitive to the world around them than one would ever imagine from their clanking clumsiness out of water.

In due course, I was to see other instances of small receptors on the surface of lobsters and crabs which responded to strain in the cuticle and effectively provided a sense of touch for an animal living in a suit of armour. Yet what was still unusual about the organs in the tips of the legs was the large size of their nerve cells and those of the joint receptors. The explanation lay in the biophysics of nerve fibres. The speed with which nerve cells conduct impulses is related to their size. The act of walking requires rapid feedback from the joints and rapid reactions to contact between the seabed and the tip of the leg. Only large nerve cells can supply the necessary quickness of reaction, an involuntary quickness I had benefited from myself in reacting to the upset crucible.

If I had gained some insights into the mechanical senses of crustaceans, I had still to get to grips with their chemical ones. I say 'chemical' advisedly because underwater it is not always possible to make a clear-cut distinction between smell and taste. We ourselves smell with our noses and taste with our mouths. Receptors embedded in our nasal membranes respond differently to a very wide variety of molecules suspended in the air we breathe. Receptors in our tongues respond much more simply to substances in our mouths to signal the sensations of bitter and sweet, salt and sour to our brains. I found no evidence for such a separation of function in the chemical senses of lobsters and crabs. Chemical sensitivity was nevertheless a property which was widely present among the setae of the claws, the legs and the mouth parts and the small antennae, the so-called antennules. The setae and their receptors showed a surprising degree of local specialization. For instance, the inside edges of the small claws at the end of each of the first two pairs of walking legs were lined with little pad-like setae. When the claw picked up a potential item of food, the pads in contact with it would bend to expose the carpet of chemically sensitive branches which formed one side of the pad. A lobster picking up an object on the seabed can therefore tell at once whether what it has grasped is worth passing on to its mouth parts. I wonder how much junk food we would eat if we could first taste it with our fingers and thumbs.

One big difference between the mechanical and the chemical senses is that, on the whole, chemo-receptors are not required to respond with lightning speed. Thus small nerve cells will suffice. This is an advantage for the chemical senses because it means that more receptors can be packed into a given space. The molecules that stimulate these senses are often highly diluted, so increasing the density of potential receptors greatly improves the chances of their being detected. However, one problem with all chemical senses, including our own, is their tendency to adapt out. In other words, if a stimulatory substance is in contact with a receptor cell for too long, the receptor ceases to respond. For an animal like a lobster, in which the chemical sense is the key to finding food at a distance, making sure that its most sensitive chemical receptors are clear of adherent food particles is vital. It was therefore rewarding to discover that there were special chemically sensitive comb hairs on one pair of mouth parts whose function it was to keep the super-sensitive chemo-receptors on the antennules clear of any particles that would have reduced their sensitivity. Effectively, when a lobster pulls its antennules through its comb-haired mouth parts, rather like a mouse cleaning its whiskers, it is tasting whether its 'nose' is clean.

By the end of my three years as a research student, I had learned to look at crustaceans with much of the admiration that I had long reserved for fishes. A great deal of their mystery remained, but at least I had had the privilege of investigating directly the information available to lobsters and crabs in conducting their lives. How much sense they made of it in terms of what we would understand as self-awareness I had no means of knowing. My conclusion at the time was that, for all the apparent complexity of their behaviour, they were still, by comparison with fishes, rather dim animals. I had seen abundant evidence that they could match their responses to the sensations they were receiving, but whether these were any more than the actions of an elaborate republic of reflexes, I could not be sure. Seen from the point of view of the genetic material whose carriers they are, whether or not lobsters or crabs know what they are doing is a matter of perfect indifference. As perpetuators of their nucleic acid puppet masters, they are supremely successful animals.

My wife, Freda, honeymooning at
Wells-next-the-Sea in 1967

Intellectually, of course, all of this playing around with the senses of lobsters and crabs was highly satisfying yet I really hated being indoors every day. Had I not made the best decision of my life in marrying a lovely local girl, Freda Carstairs, whose family had lived in the town since the fifteenth century, I am sure I would have quit in despair and sought a job as a gamekeeper or fisherman. Eventually, though, the time came when I felt that I had done enough on the practical side to write up my results and submit them as a thesis to the examiners. It was not as grim a task as I feared because the editor of a scientific journal had published the results of my early work and I had another paper in preparation. I was, however, dreading the *viva voce* with the external examiner, the tall and taciturn J. W .S. Pringle, Linacre Professor of Zoology at Oxford. There he was known by undergraduates as 'the smiling don' in ironic recognition of the austerity of his demeanour. I was shortly to face his interrogation.

For some reason, the meeting took place in Glasgow in a building the vestibule of which was covered from floor to ceiling with white tiles. I remember wondering why this was. On the way in to Glasgow I had seen a number of tenement walls defaced by graphic references to the alleged shortcomings of His Holiness the Pope. Perhaps the tiles were there to guard against such artistry? Whatever they were for, their appearance was sanitary rather than welcoming and did nothing to lower my heart rate.

I stumbled up in a daze to the condemned cell, a cheerless little room overlooking a row of fly-blown dustbins at the back of the university's physiology department. Pringle offered his hand and then, with a smile, put me out of my misery at once, the rest of the interview being confined to points of detail and parallels with his own work in the 1930s, when he was one of the pioneers of the Cambridge tradition of experimental biology. In this, my only experience of him, Pringle was anything but the glum soul he was reputed to be.

THE GREAT M^CINTOSH

That the Gatty Marine Laboratory existed at all, and that I had been able to study lobsters and crabs there in such detail, was entirely thanks to the remarkable personality of its founder, William Carmichael M'Intosh (usually spelt McIntosh), Professor of Natural History at the University of St Andrews from 1882 to 1917 and a giant figure in the early history of British fishery research.

McIntosh was a local lad, born in St Andrews in the autumn of 1838, the only son among three older sisters and two younger. McIntosh is not a lowland name and, although William's father John was born in Edinburgh and had lived from infanthood in St Andrews, his great-grandfather was from Inverness-shire. John was a self-made builder who rose to prominence in St Andrews during its period of transition from a sleepy relic of the Scottish Renaissance and Reformation to a graciously appointed centre of learning, business and golf.

In many respects, McIntosh was the quintessential east-coast Scotsman. Unlike its southern neighbour, Scotland took the Reformation seriously and many of what are still thought of today as the cardinal virtues of the Scots character – thrift, hard work, attention to detail, respect for education and a dislike of outward show – owe their origins to the Bible-worshipping divines of Geneva. At the time of his marriage to Eliza Mitchell of St Andrews, John McIntosh was a member of the Established Church. This was not and is not the Scottish Episcopal Church but the Church of Scotland, a sternly Presbyterian organization, loyal to

Professor William Carmichael McIntosh
studying fish larvae in his laboratory

the Sovereign but acknowledging no head but the person of Jesus Christ himself. Eliza, who was to become the single most important influence on the life of her only son, was a devout member of the Congregational Church in St Andrews, another even more democratically organized foundation which was then enjoying the preaching of the talented theologian and scholar, the Reverend William Lothian. Young William McIntosh was taken to hear Lothian's preaching on many occasions and the memory of this ministry was to remain with him for the whole of his long life. The influence of both his mother and Lothian's biblical teaching were reinforced by his attendance as a young child at Miss Bell's Congregational Sunday School, an innovation for St Andrews that most boys would surely have found thoroughly irksome, even if William did not.

After a brief spell at infant school, William McIntosh entered Madras College, St Andrews. He remained there until he was fifteen, studying the full range of subjects including Latin and physical geography, the latter being the closest the school got to the teaching of practical science. There is no doubt that William enjoyed his schooldays. He was profoundly interested in the world around him and set those who were able to enlighten him on something of a pedestal. Thus, writing of his mathematics master, Dr William Oughterlonie, he recalls that he

> ... was worthy of all praise both for the perfect order of his
> classes and for his ability ... He also taught the various classes of
> geography in a manner which elicited the enthusiasm of the
> intelligent pupils. Our text-book was Rhind's *Physical Geography*,
> a little work in two parts, which gave a charming view of natural
> objects. I remember the delight felt in becoming acquainted
> with the distribution of animals, and in knowing the clouds. Dr
> 'Lonie was a little man, and one eye had been seriously injured,
> but he was active and vigorous.

William took a strong interest in microscopical exhibitions organized for the public by the university and was well aware of the work of Professor John Reid, a pioneer of marine biological study in Scotland who was sadly to die of buccal cancer at the age of forty. In McIntosh's own words, '[Reid] died of cancer of the tongue after two operations [in pre-chloroform days] which he heroically bore without a murmur, even smiling as Professor Syme's knife severed certain tissues. Such heroism is rare.' We can be quite certain that McIntosh saw nothing ironic about that last sentence.

It was not, however, until the third session of his attendance at St Andrews University that McIntosh received formal teaching in the biological sciences. Reid's successor as Professor of Medicine and Anatomy was the equally outstanding George Edward Day, a contemporary and friend of the prominent nineteenth-century naturalist Edward Forbes and the anatomist John Goodsir. McIntosh was to learn a great deal from Professor Day, whose lectures covered both physiology and comparative anatomy. As a result of his teaching and that of the Professor of Natural History, William MacDonald, McIntosh single-mindedly pursued the goal of seeking a qualification in medicine, not so much out of a desire to heal the sick as because it was the only means then available to study the animal world without the support of a private income. He was admitted to Edinburgh University's prestigious Medical School in 1857, having spent the summer vacation since leaving St Andrews at his father's small rented farm, Sunnybraes, which lay not far from the town on the road to Largo; it was Sunnybraes to which McIntosh was to owe his lifelong love for all things rural and his habit of rising at five o'clock in the morning.

McIntosh's entry to Edinburgh University's School of Medicine coincided with one of its several periods of world pre-eminence and he enjoyed instruction from such medical giants as Joseph Lister, the pioneer of antiseptic surgery, James Young Simpson, the discoverer of chloroform anaesthesia, and John Goodsir, the progenitor of the theory that the cell was the fundamental unit of what was then called the vital process. It was to be Goodsir who exerted the strongest influence on the young McIntosh. By inclination a naturalist rather than a clinician, rightly asserting that 'No great human anatomist ever lived who was not a great comparative anatomist', Goodsir had heard from George Edward Day about the promising young man who was shortly to join his classes and took him under his wing.

McIntosh also benefited from George James Allman's course of fifty lectures in natural history. Allman was a native of Cork who is remembered today in the specific name of the less common of the two species of brown shrimp, *Crangon allmani* Kinahan. He was a kindred spirit who took McIntosh on his first dredging expedition, exposing him in the process to that bane of the casual seafarer's life, *mal de mer*. The two were to remain in touch until Allman's death in 1898, and it is conceivable that Allman's consummate draughtsmanship was the inspiration for the beautifully illustrated monographs on invertebrates which, completed with the help of his artistic sister Roberta, were to be the crowning glory of McIntosh's career as a naturalist.

McIntosh applied himself with equal fervour to his medical studies, living a strangely cloistered life under his austere precept that 'No man who joined bonhomie to wine and women could qualify as a true votary of science'. He graduated MD as a gold medallist in 1860, having completed a notable thesis on the nervous system and behaviour of the shore crab, *Carcinus maenas* L., for Goodsir's prize essay. These crabs are lively and aggressive masters of the world between the tides, and so complex is their behavioural repertoire that it can be mistaken for intelligence. McIntosh was one of the first British naturalists to show by direct experiment that much of what appears to be purposeful activity by a self-aware organism is more easily explained as the well-orchestrated stringing together of relatively simple reflexes. His method was to combine a comprehensive understanding of the detailed anatomy of the shore crab with careful observations on the responses of the crab to simple stimuli. For its day, it was a fine piece of work, but no doubt his use of strong acids to determine the 'irritability' of his hapless victims would no longer be encouraged.

Although McIntosh's primary aim was to establish a reputation as a naturalist, he was determined to do so from a position of financial strength. He was especially concerned not to depend on his father, to whom he already felt greatly indebted. He accepted the post of Assistant Physician at Murray Royal Hospital in Perth where the Physician Superintendent was a keen botanist. The hospital, which was a mental institution, had been equipped with a natural history museum where the minds of at least some of the poor disturbed patients were to benefit from the new Assistant Physician's zoological lectures.

A new mental hospital was opened at Murthly in Perthshire in 1863, and McIntosh was appointed its Physician Superintendent, once again seeking to combine his medical duties with his zoological interests. He held the post for nineteen years and, although he had a few problems with the Lunacy Commission who were concerned that the mortality rate at Murthly was unduly high, his successes as a naturalist secured him Fellowships to the Royal Societies of both Edinburgh and London. Over the same period, he also took part in the analysis of dredged material from the *Porcupine* and *Challenger* expeditions, endeavours which marked the British Empire's coming of age as a world leader in the new science of oceanography. Exact description and illustration is the essential first step in any new area of biological study, and no one was better at it than McIntosh. His attention to detail and his ability to draw exactly what he saw were so good that his material is still used over a century later by ecologists studying the deep sea.

Given his solid record of achievement, it is perhaps surprising that McIntosh was not to gain an academic position until 1882, when he was appointed to the Chair of Civil and Natural History in his old university of St Andrews and was forty-four years old. Four disappointments had preceded this happy development, and one can only speculate about the reasons for them. A lunatic asylum is not a promising address from which to seek entry to the higher echelons of academic life, and it could well be also that the somewhat cheerless determination which had enabled McIntosh to pursue two careers in a sort of double harness had made him rather forbidding at interview. Interestingly, his rejection by Aberdeen in 1878 was said to be because of his alleged reputation as a 'hot Darwinian', a posture which at that time was anathema to both McIntosh himself and the ecclesiastical authorities in that university. But within a year of his appointment to St Andrews, McIntosh was recruited to the Trawling Commission under Lord Dalhousie, an organization of which the great biologist Thomas Huxley was also a distinguished member. Huxley sent McIntosh the following letter:

> You are probably aware that I am a member of a Royal
> Commission … which is inquiring into the grievances of the
> Line fishermen against the trawlers. We find it very desirable that
> some more precise and specific information should be obtained
> as to the action of the trawl-net in bringing up 'spawn' and
> immature fish; and for this purpose it is probable that we shall
> want the services of a scientific naturalist, who will have the
> means provided of trawling over some piece of ground (to be
> hereafter determined) at regular intervals for the next six months
> or so, just as a trawler would do, and whose business it will be to
> register the nature of the products of each haul of the trawl. Lord

Dalhousie wishes me to ascertain whether, supposing this plan
to be carried out, you would be inclined to undertake the work.
A proper vessel will be provided. Expenses will of course be
paid, and there will be an honorarium, but I can say nothing
definite on this head until the Treasury has sanctioned the plan.
In fact you must understand that the whole affair is as yet
problematical, but I believe the plan will be carried out, and I
am now anxious to know who can be got to undertake the
work. I imagine the ground chosen will be off the east coast of
Scotland, and that it will not be necessary to trawl more than
once a week. Please let me have your reply as soon as you
conveniently can …

William McIntosh did not need to be asked twice.

TRAWLING FOR TROUBLE

By the 1880s, the British fishing fleet was at something of a cross-
roads. Increasing numbers of vessels now plied home waters to
meet the great demand for fish from the growing urban population. Although the small yards around the coast that specialized in
building fishing boats had introduced some innovations to improve
their sea-keeping qualities, most of the new vessels were still
powered by sail, and many still used fishing gears which depended
for their effectiveness on active responses from the fish to swim
into walls of net or take baits and for the shellfish to crawl into
traps. As a result of this conservatism in both the vessels and their

*A sailing smack towing a beam trawl
in the North Sea*

equipment, the total capacity of the increasing fleet to kill fish was still small by the standards of today. Even when the effects of fishing on the abundance of the stocks had had a noticeable effect on catch-rates, it was usually possible to restore them by fishing farther from port.

In retrospect, it is astonishing that it took so long for steam to find its way into the engine rooms of the fishing fleet. Towing dredges for molluscan shellfish like oysters and primitive beam trawls for flatfish like plaice and sole had been practised by sailing smacks for many years. However, towing a gear liable to snag rocky outcrops on the seabed with a vessel whose course is controlled as much by the direction of the wind as by the need to avoid obstructions greatly restricts the efficiency of this method of fishing. So, rewarding as trawl-fishing by sail could be, its inherent limitations prevented it from ever presenting a serious threat to the stocks. All this was to change with the introduction of steam to the fishery, first by using a steam tug to tow the smacks and their trawls and later by fitting the smacks with steam engines of their own.

The introduction of mechanization to fishing the seabed, along with improvements in the design of trawl nets to improve their ability to herd fish into their path, caused great alarm among inshore fishermen using traditional methods. Although part of their concern might have been that the steam trawlers would take fish they would have preferred to catch themselves, their appeal to the authorities was made on the supposition that trawl-fishing, especially by steam, would 'impoverish the fisheries' and, as Huxley's letter indicated, bring up 'spawn and immature fish'. Biologically speaking, the fishermen's second point was the more worrying of their objections in that it concerned the capacity of successive generations of the stocks to replenish themselves. It was, however, open to direct investigation. McIntosh's response, on taking up his appointment, was to put in hand an immediate programme of practical work at sea, accompanied in the early work by Lord Dalhousie himself. He supplemented his marine programme with direct observations on the reproductive products of captive fish in the former fever hospital at St Andrews which was recommissioned as a Fisheries Laboratory using funds provided by the Fishery Board for Scotland.

It became clear early on in the scientific programme that the eggs and earliest life-stages of most of the fishes like haddock, cod, whiting and lemon sole, whose offspring were thought to be at risk from bottom trawling, lived in mid-water. They did so alongside the host of other fish and shellfish larvae and their food organisms that scientists collectively call zooplankton. This news was not well-received by the line fishermen, who had previously been strong supporters of McIntosh's investigations. Nevertheless, McIntosh made no secret of his conviction, based on what he had seen so far of the fish, their eggs and the practice of trawling, that there was no case for closing the inshore grounds to this new form of fishing. This was too much for the line fishers, whose muttered threats culminated in a noisy demonstration outside McIntosh's house during which a tarred effigy of him was set ablaze. For the first time in British experience, scientist and fisherman found themselves on opposite sides. It was not to be the last.

McIntosh's effigy burned because his biological answer to the fishermen's ostensibly biological question did not address their immediate fear – namely that, whether or not the fish stocks themselves were at risk, their share of them certainly would be. Nowadays we would describe the problem as largely one of resource allocation, and the call for conservation measures for reasons which were essentially selfish has a decidedly modern ring to it. In another sense, we can only marvel that, at a time when fishing pressure was but a fraction of today's, McIntosh was able to provide valid advice on a stock management problem without any understanding of the capacity of the resource to withstand additional exploitation.

When McIntosh began his work in support of Lord Dalhousie's Royal Commission, he was faced with something of a blank canvas. Despite the world pre-eminence of the Royal Navy and the comparative prosperity of the British fishing industry, the nation's understanding of the biology of the principal food fishes had fallen well behind that of Norway and even the United States of America. Surprisingly, in view of their earlier hostility, one of the main sources of new material for McIntosh's studies was submitted by the fishermen themselves. The catalyst for this conciliatory development was a young man called Edward Prince, an eminent scientist in his own right with degrees from both St Andrews and Cambridge. Unlike McIntosh, he had a warmth about him that appealed to the fishermen and it was not long before even some of the more antagonistic were borrowing jars from Prince and bringing in fish eggs they had recovered from surface waters during their normal fishing operations.

In attaching Prince to McIntosh's laboratory, the Fishery Board for Scotland had secured a double benefit: a loyal colleague of the highest academic quality for their principal investigator and the renewed support of the fishing industry. Such a combination promised much. By 1888, the life cycles of forty or so food and other fishes had been worked out. McIntosh had reported and illustrated the results in a range of publications and the time had come for a thorough stock-taking. This he accomplished with the help of Prince in a contribution to the *Transactions of the Royal Society of Edinburgh* which ran to over 270 pages. The collaboration with Prince, who in 1892 was appointed Commissioner of Fisheries for Canada, marked the personal high point of McIntosh's career as a fishery scientist. However, his influence continued long afterwards through the work of the many distinguished scientists from England, Norway and elsewhere in Scotland who came to work with him in St Andrews.

By the mid-1890s, it was time to put the laboratory facilities at St Andrews on a more permanent footing. McIntosh secured the generous support of Charles Henry Gatty, a gentleman of independent means, who was a fellow member of the Ray Society. Gatty not only paid for the construction of a new laboratory, 'the style of which is a simple treatment of the English Renaissance', but also for its complete equipment. The new Gatty Marine Laboratory was opened in 1896 and McIntosh and his collaborators were to make full use of it until the outbreak of war in 1914 took its inevitable toll on the size of the workforce.

If McIntosh's personal contribution to the development of fishery science in Great Britain had been considerable, the activities of the Fishery Board for Scotland had been less productive. Part of the Board's problem lay in the tension which developed between the requirements of political expediency and the more clean-cut guidance provided by scientific advice, especially that of McIntosh. He had not been best pleased that, in the days of Dalhousie's Trawling Commission, areas had been closed to this form of fishing in the face of his evidence that the reproduction of the principal food fishes was not under threat. Similar disagreements developed with the Fishery Board, which, in McIntosh's eyes, had further compromised its position by relying on field results collected from the *Garland*, a former pleasure yacht entirely unrepresentative of the fishing fleet. In the absence of the wise conciliatory counsel of Lord Dalhousie, who had died, relations between the Board and McIntosh slowly deteriorated. His final advice to them was that, in judging the effects of fishing on the stocks, they needed to consider the combined effect of trawlers and liners together; in other words, to look at the problem from the point of view of the fish which were killed just as surely by one fishing method as another.

Not all of McIntosh's advice was so sensible. Along with Thomas Huxley, he was persuaded that nothing Man could do with the fishing methods then available could possibly impoverish the seas.

This concept he developed in his famous popular book, *The Resources of the Sea*. Huxley had come to the same conclusion from first principles. McIntosh gave equal weight to Divine Providence. The problem was that the steam trawlers of the 1890s were very different performers from their predecessors of the 1880s. It soon became clear that, on the more heavily fished grounds, both the catch-rates and the average sizes of the fish were not what they used to be. McIntosh refused to change his views and this intransigence, along with his justifiable criticisms of the Board's practical work, may have cost the Gatty Marine Laboratory its financial support. For many years the laboratory would be sustained by McIntosh's programme of marine biological studies but, after his retirement from the Chair of Civil and Natural History in 1917, his successor, the flamboyant Anglo-Irish polymath D'Arcy Wentworth Thompson, had nothing to do with the Gatty.

Fortunately, McIntosh was made of sterner stuff and continued to work there, insulating his feet in a box filled with straw during the long winters. He also participated in the discussion of fishery issues when the occasion arose but, as the years went on, the untenability of his point of view as he then expressed it became increasingly clear to new generations of fishery scientists. Indeed, after McIntosh's death in 1931, the Gatty Marine Laboratory passed into suspended animation. Evidence that it had not been entirely forgotten by higher authority was proved in January 1946, when storm-force winds from the North-East swept away part of the sea defences in front of it. Remedial work began the following April, and not long afterwards, McIntosh's successor, D'Arcy Thompson, appointed the endocrinologist Dr Jimmy Dodd as Lecturer-in-Charge. Under Dodd's stimulating direction, the laboratory came back to life, even acquiring a white-painted but none too seawor-thy vessel, the *Argonaut*, for work in St Andrews Bay. Such was Dodd's success in rebuilding the laboratory's scientific reputation that he was able to attract sufficient funds to extend it both for his own work and for that of the other zoologists and botanists who worked with him.

Dodd was succeeded as director by Adrian Horridge, a disciple of Carl Pantin of Cambridge who, in a Reith Lecture, made the startling pronouncement that the function of biological science was to enable Man 'to predict and control the biological future of the earth' – not, perhaps, a sentiment that would have appealed to McIntosh. Pantin's efforts to predict and control his protégé were somewhat desultory, however, and as a research student Horridge was largely left to work out his own destiny. On his own initiative, he transferred his attention to the study of the ostensibly simple nervous systems of anemones, corals and jellyfish. This early work and his collaboration with an American scientist, Ted Bullock, in

McIntosh's drawings of life stages of the shanny,
Blennius pholis *L.*

the writing of the magisterial neuro-biological work, *Structure and Function in the Nervous Systems of Invertebrates*, established his reputation. And, of course, it had been Horridge's newly appointed colleague, Mike Laverack, who had reacquainted me with lobsters.

As for McIntosh himself, he was to be paid a gracious tribute by Michael Graham, the outstanding director of the Ministry of Agriculture, Fisheries and Food's Lowestoft laboratory in the middle years of the twentieth century. In Graham's words, McIntosh had been 'entirely wrong but so nearly right'. What Graham meant was that, although there was no doubt about the capacity of the fishing fleets to impoverish yields in terms of weight, there were rather few cases at that time when the supply of young fish had been put at risk because of fishing and no case where fishing alone had rendered a species of fish extinct.

Perhaps McIntosh's greatest contribution to the science of fisheries had been to lay the foundation for our knowledge of the life cycles of the main food species. It was work that required great persistence and stubbornness in the face of adversity, and meticulous attention to detail. Much of his original material, beautifully preserved and labelled, still exists in St Andrews and Aberdeen, along with his original drawings and notes. Only the most single-minded of investigators could have left such a rich legacy. But, capable as McIntosh was in this descriptive work, his main limitation was his inability to think in a more dynamic way about the forces that lay behind the growth, mortality and reproduction of the species he studied. Ironically, it is conceivable that he would have made greater progress in this area of understanding had he listened more to the fishermen, the very people whose concern had been the stimulus for his life's work.

A GRAIL OF SORTS

After so much time spent in the study of crustacean toenails, I was ready to resume my quest for the Grail, in the form of a decent eight-bore. As it was, and rather unexpectedly, the Grail, if such it could be called, came to me.

One of my brother Peter's friends from schooldays was called Backhouse. He had lost his father early in his life but he had a most spirited mother who sought to make up for this loss by encouraging her son in every possible sporting activity. On one occasion, Backhouse's mother took my brother Peter, her son and me to Hurley on the Thames where, Backhouse alleged, it was possible to catch bleak, *Alburnus alburnus* (L.), with blowfly maggots or gentles at the rate of a fish a minute.

Bleak are small shoaling fish of the carp family and they are specially adapted for life at the surface, with a mouth upturned to seize the smallest of floating insects. They are narrow, pointy-headed fishes which, looked at directly from above, appear a beautiful shade of sage green. From any other angle, they are the brightest silver, a

Bleak, Alburnus alburnus *(L.), and other boys' favourites*

property made use of in the past in the manufacture of artificial pearls by dipping glass beads into a silvery brew called essence of orient. The essence was made from a concentrated infusion of bleak scales. This infusion was left long enough for the guanine crystals, that make bleak look silvery, to separate from the outer faces of the scales, which were then sieved out. Whether or not the resulting mercurial liquor contained enough fish protein to stick the guanine to the glass beads, I do not know. I suspect that, had the liquor been allowed to concentrate by evaporation, it would have done so. The resulting beads were very attractive, if a little flashy.

We travel to Hurley in the Morris Traveller shooting brake belonging to Backhouse's mother. In the back is our fishing tackle, the tins of gentles and rather an elaborate picnic. It is a glorious summer's day, unpromising for most forms of fishing but not for bleak. They are bright spirits which sport in the surface, the vastness of their shoals itself a protection against attack from above or below. We cannot wait to tackle up, fumbling to knot the catgut casts with their small floats on to the lines on our reels. Backhouse advises setting the gear to fish shallow and putting on just enough split shot to cock the float. Out into the stream swing three sets of tackle, three floats disappear at once and three gleaming bleak are unhooked. This is unbelievable to Peter and me, accustomed as we are to spending a whole day on the Aylesbury arm of the Grand Union Canal with two perch and a small roach our only catch. As usual, I get into a tangle, but Peter patiently sorts it out and the sport is resumed.

It has been said that, for those who pursue it long enough, an angling career has three phases: the desires to catch the most fish, the largest fish and the most difficult fish. There is no doubt which stage we were at that morning at Hurley. The bleak are beginning to accumulate in the keepnets, but we are getting worried about the supply of gentles used as bait. Our tin is upset during the initial frenzy, fortunately lower down the bank than the picnic baskets. We try to recover as many as we can, but large numbers escape to freedom among the grass stems. The second tin still has a fair amount in it, but we try and make one gentle last for several fish, even if all that is left on the hook looks like an empty sausage skin in miniature. The bleak do not seem to mind and, by the end of the afternoon, our anglers' spring balances, which we have never had cause to use before, tell us that we have caught over six and a half pounds of fish. What are we going to do with them all? Backhouse is unsure, but for Peter and me there can be only one answer – eat them, of course.

When we got home, still highly excited, we found that our catch totalled 127 fish, a little short of the 153 of the miraculous draught recorded in the Gospel story, but still a highly satisfactory result. Heading and gutting the bleak took some time and they then had to be transferred to a large bowl of brine, with a little malt vinegar, to remove any hint of earthiness. It took three days to eat all of the fish, tossed in seasoned flour and fried over a brisk heat, a heat so fierce that at one stage a pall of acrid smoke filled the kitchen. By the third day we were eating them cold, still with enjoyment but reaching the point at which a return to my mother's cooking would have been welcome. By the end of that evening, we had eaten all our 127. Over forty years later, I think I could just about face the snow white flesh of another one.

Backhouse's conversation was highly colourful. He seemed to live in a sort of fantasy world in which the line between imagination and reality was more than a little blurred. If at times some of his stories tested our credulity, the miracle of Hurley did much to restore our faith. He also spoke of his shooting experiences in the beechwoods between Chesham Bois and Amersham, public areas not normally associated with the sport of St Hubert. Allegedly, they used to proceed in single file, an old gaffer with a twelve-bore hammer gun in the van, Backhouse next with a single-barrelled, muzzle-loading ten-bore and a tail of smaller boys, variously armed with air rifles, air pistols and catapults. Woodpigeons, jays and grey squirrels were said to be the main objects of the chase but, by all accounts, bags were light.

At the time, I was never quite sure how much of this story was really true. But just as the discovery of the Dead Sea Scrolls shed new light on the origins of Christianity, so the discovery of Backhouse's ten-bore – the nipple, the part on which the percussion cap sits so that it can ignite the powder charge when the trigger is pulled, by this stage rusted solid – added credence to his accounts of the feats of arms among the beech trees.

By now, Backhouse had been banished by his mother to relatives in Bolton. She was worried by his association with a mystic cousin who sought spiritual enlightenment through the study of partially fogged black-and-white photographs which he had printed himself. Backhouse's ten-bore had been left behind with the mystic, who thereupon sold it to my father with some old guns of his own. These included a long-chambered eight-bore made by the firm of Henry Holland, later to become the renowned Holland & Holland, in the late 1860s. It fired green cartridges about four inches long. In its day, it would have been a formidable performer on the shore. Unfortunately, as so often used to happen with these heavy old guns, its barrels had been shortened, which meant it shot barn-door patterns all over the landscape and no longer allowed the black powder in the long green cartridges to develop their full power.

I was nevertheless glad to have this battered Trojan in my hands as I waited in the marram grass fringing the Eden estuary in the hard winter of 1963. As the sun came up, the widgeon started to move, the drakes making their wild whistling call. The activity seemed to be confined to the middle of the estuary and all I could do was watch the spectacle as parties of widgeon, long-winged for such small duck, moved between the mud banks. I was glad enough to be there to drink in the wildness of it all, even though the 'fowl were far out of range and the temperature had fallen further with the dawn. I started to shudder with the cold. The flight petered out and, getting creakily to my feet, I walked along the narrow fringe of salt marsh which separated the mud from the sand dunes.

'Whee-o, whee-o' – widgeon on the wing and very close. They skimmed over the marram-grass tops so quickly that I only just had time to cock the hammers of the eight-bore. I brought it clumsily to my shoulder and fired at the leading bird. There was a colossal explosion and both barrels went off almost simultaneously – the sears of both locks were badly worn – and I ended up on the ground. Needless to say, I did not touch a feather. For some reason, possibly because I was a little disorientated, I decided to stay just a little longer. Another party of widgeon powered over the sea grass and this time I cocked only one hammer. Another big bang, an orange flame accompanied by a lot of grey sulphurous smoke and two plump widgeon, both drakes, lay breast uppermost behind me. It was to be the last hurrah of the old gun, at least in my hands.

Some years later my dear wife, who has a better head for commerce than I have, sold it for £45, the same price as my father's incomparable Purdey.

Another venerable gun was to be in my hands on that red-letter day in the lives of all young wildfowlers, when they shoot their first goose in the dawn. The goose was to fall to a hammerless ten-bore built in the mid-1880s by that great maker of wildfowling guns, J. & W. Tolley of Birmingham and London. My father had bought the gun before the Second World War for Joe Ballard, an old farmer who had been a contemporary of my Shelton grandfather. Old Joe occasionally used to accompany my father on his winter expeditions to Wells-next-the-Sea, but sadly no goose fell to his ten-bore. Following his death, my father bought the gun back, and it was now with me in Fife.

My father had provided me with the last of his three-and-a-quarter-inch ten-bore cartridges, hand-loaded with Schultz smokeless or black gunpowder. Schultz gunpowder was named after Captain Ernst Schultz, an officer in the army of Bismarck's Prussia. It was the first gunpowder not to produce a copious and evil-smelling smoke when ignited. The snag was that its performance was rather sensitive to the tightness of the turnover, the rolling over of the end of the cartridge case which retains the top wad – and therefore the contents of the case – in position. The black gunpowder was unaffected in this way and gave a consistent performance, albeit at the expense of smoke which, on a still day, could obscure the target for a second shot. The trick, therefore, was to have a Schultz cartridge in the right barrel for the first shot and a black powder one in the left for the second, should it be required. It often was. As my father's old wildfowling friend, the Reverend

Walker, Episcopal Rector of Castle Douglas in Kirkcudbrightshire, used to say: 'Miss with the smokeless but kill with the black.' (The good priest knew whereof he spoke: he would further improve his chances on snowy mornings by wearing his surplice over his 'fowling smock.)

My father-in-law and I had motored out of St Andrews on the Largo road to some fields owned by his friend Robert Graham near Cameron Loch, a small body of water which was popular as a roost for pinkfoot geese. We were there well before dawn so had rather a long wait before the geese started to move. We could nevertheless hear them calling to one another out on the loch – to a wildfowler their music is the most exciting sound in the world. They had arrived in east Fife in early October, having flown down from their summer breeding quarters in Iceland. There were probably a few of the later-arriving greylag geese there as well, but we could not distinguish their more guttural calls against the increasing noise of the pinkfeet. At last the chorus reached a crescendo; we heard the sound of wings and feet striking the water and the first party took off.

I slip the ten-bore out of its cover, fumble for my cartridges, smokeless in my right-hand pocket, black in my left, and both loaded with BB shot. A glance down the barrels to check that they are clear and the cartridges are in the breech, ready to do their duty. The geese fly out over the other side of the loch followed shortly after by two more parties of their fellows. There is nothing we can do about it. The fields on the other side belong to another farmer, and we have permission only from Robert Graham.

As so often with 'fowling, there is much pleasure to be had from just being here, enjoying sights and sounds denied even to most birdwatchers. Geese are rather late risers and, long before the first parties take off, we are entertained and not a little tempted by the whistling sound of mallard passing over in the half-light. The early skeins of pinkfoot merely entertain. They are far too high for even a tyro like me to chance a shot. As the waving lines pass overhead, the small heads of the geese thrust forward on necks longer than those of the mallard, it is difficult to believe that they are probably already travelling at almost fifty miles per hour. They have not been flying long enough for there to be sufficient separation among the birds to create many vee-formations; their current imperative is to gain height and to make for their grassy feeding grounds with a minimum of delay. I cannot resist swinging the long barrels of the Tolley through one of the leaders. The entire bird, including the wing-tips, is blanked out – proof positive that it is well out of range.

'The hounds of heaven'

It is getting really light by now and we have to take more care in concealing ourselves among the sprots or rushes near the edge of the field. 'Wink-wink', 'Wink-wink' – 'Where the hell was that?' It is certainly louder than any of the other geese. Suddenly we both see it, coming straight at us and only about thirty yards up. This is no time for the uncertainties of smokeless powder. Up comes the ten-bore just as the single goose passes overhead. There is the usual orange flash and then the grey smoke which hangs in the still air. A heavy thump behind and all is well. I have slain the Jabberwock and am feeling pretty beamish about it.

AN ILL WIND

At around nine o'clock on the morning of Saturday, 18 March 1967, the tanker *Torrey Canyon* ran on to the Pollard Rock, part of the notoriously dangerous Seven Stones Reef off Land's End in Cornwall. The ship was fully laden with topped crude oil from Kuwait and was making its way to the Milford Haven refinery complex to discharge its cargo. Six of the ship's eighteen oil storage tanks were torn open by the initial seventeen-knot impact. There was no realistic prospect of refloating the ship and by 26 March she broke her back in the heavy seas pounding the reef. Attempts to set fire to the oil, which included bombing attacks by Buccaneer strike aircraft of the Fleet Air Arm, succeeded only in igniting the more volatile fractions of the petroleum. By this stage, the bulk of the 117,000-ton cargo was out of the ship and mixing with the turbulent sea to form a sticky water-in-oil emulsion which the press was quick to christen 'chocolate mousse'.

The economic priority for the Government of the day was to get rid of as much as possible of this intractable material before the start of the holiday season. This the authorities attempted to do by collecting all that they could mechanically and by treating the rest with dispersant mixtures known as solvent emulsifiers. It was not

The tanker Torrey Canyon *lying broken-backed
on the Seven Stones Reef*

long before those with knowledge and imagination, and a much larger company of those who possessed neither, began to question the clean-up policy, fearing that the treatment was making a bad ecological problem worse.

The Ministry of Agriculture, Fisheries and Food was involved from the beginning through the able participation in the planning and monitoring process of Arthur Simpson, head of its laboratory at Burnham-on-Crouch in Essex. Rightly, Simpson was to conclude that, such was the diluting power of the strong tides of the western Channel, the effects on fisheries amounted to little more than the destruction of some eggs of the pilchard, *Sardina pilchardus* (Walbaum), in the open sea and the tainting of lobsters in holding boxes floating in some of the Cornish coves. Onshore, the problems were much worse. Concentrations of ten parts per million or less of the dispersant mixtures could be toxic to marine animals and plants, and the local effects were catastrophic, especially on rock-living herbivores such as limpets and filter-feeding barnacles. These losses, and recovery from them over the succeeding years, were duly studied, and the results published, by the staff of the Marine Biological Association's Laboratory at Plymouth.

Public reaction to the disaster was centred initially on the tragic fate of the sea birds, mainly shags, guillemots, razorbills, puffins and a few great northern divers. Some 25,000 of these birds were thought to have died from hypothermia and the effects of swallowing oil while vainly attempting to preen it from their feathers. As time passed, so public attention shifted to the ecological problems highlighted by the work of the Plymouth laboratory. Back in Lowestoft, Dr Herbert Aubrey Cole, the director of Fisheries Research for England and Wales and of whose organization Arthur Simpson's unit was a part, saw his opportunity. Privately, he was greatly relieved that English fisheries had escaped virtually unscathed. In the corridors of the Ministry's headquarters in London, however, he took a very different line.

For some years Cole had been trying to build up a specialist pollution research group at Burnham-on-Crouch. He had already appointed a bacteriologist and an analytical chemist. Now perhaps he could persuade the Fisheries Secretary, the urbane Cambridge mathematician Ian Graham, that it was time to appoint a biologist in support of Arthur Simpson. Graham was the gentlest of souls in the widest sense of the word, a public servant of the highest calibre, utterly different from the execrable spin doctors favoured by the present generation of politicians. His secretary, an older lady of great charm and firmness, always referred to him as 'Master', and we can be sure that Cole would have been ushered into the great man's room by the invariable 'Master will see you now'.

Interviews with Graham were always enjoyable – like Cole, he was an enthusiast and he had a natural ease of manner which concealed the sharpness of his brain from the unwary. Cole mustered his arguments. We had got away with it this time, but what if the accident had happened in the Thames Estuary? Had not the Prime Minister himself taken a personal interest? (Harold Wilson, who normally took his holidays in the Scilly Isles, had been photographed sitting in a helicopter with his leather flying helmet on back to front.)

Graham was unmoved by these and Cole's other wild swings. Then came Cole's master-stroke: was there not a danger that, unless the Ministry increased its commitment to marine pollution research, its influence within the Cabinet Office would be usurped by that of another department, even perhaps the Department of the Environment that Harold Wilson was rumoured to be setting up? Graham thought for a moment, drew the interview to its customary courteous end and Cole returned by train to Lowestoft convinced that, for once, he had put one over on Graham. It was to be some time before an envelope containing a letter on the sky-blue notepaper then used by the higher echelons of the Civil Service landed on Cole's desk. When it did, it was to indicate Graham's agreement in principle to the new appointment, but it was to take many months before the Civil Service Commission placed an advertisement in the scientific periodical *Nature* for a post at the MAFF fisheries laboratory at Burnham-on-Crouch.

As the end of my time at the Gatty approached, I had begun to take a closer interest in the job advertisements at the back of journals such as *Nature*. I had even responded to a few, invariably receiving the same unhelpful reply, and I was not terribly optimistic about my chances of getting the job at Burnham. I did not know a great deal about pollution, and I was not sure that I wanted to live in Essex. I sent away for the forms, completing them in rather a hurry. Back in the post they went and I heard nothing for weeks. In fact, I had long forgotten about the application when, one morning, a brown envelope arrived with the words 'On Her Majesty's Service' along its top edge. I thought that it was from the tax office but opened it anyway. To my absolute astonishment, I had been invited to an interview in London. Why on earth could they possibly want to consider me? And then I remembered; for the first time, I had included the name of J. W .S. Pringle, the 'smiling don', among my list of referees.

The interview, at the London headquarters of the Civil Service Commission, was chaired by a recently retired scientist who, like Ian Graham, was an urbane figure with all of the old-world charm commonly found in the best scientific circles before the Second World War. On one side, wearing a dreadful dark brown pin-striped suit and otherwise looking exactly like His late Majesty King Edward VII, was Dr Cole. On the other was Arthur Simpson and, beside him, an earnest but kindly-looking Welshman from the Water Pollution Research Laboratory. I have no recollection of what was actually said, only that everyone, from the chairman to the Welshman, was extremely nice. Afterwards, my wife bought me a sprig of white heather from a poor Glaswegian who clearly found London as alien as we did. We had lunch at a Lyons Corner House and made our way to a tube station. As so often on the Circle line, it was standing room only. Who should be standing next to me but the tall, languid figure of the chairman of the interview board? 'They are going to offer you the job, you know' was his welcome greeting.

A BEGGAR'S MANTLE
FRINGED WITH GOLD

To Thomas Huxley and William McIntosh, fisheries appeared sustainable because the fraction of the resource represented by catches was so small that it fell within the range of natural variations in abundance. For Huxley, the continued rise in the British catch as the nineteenth century entered its last decades was evidence enough. For McIntosh, despite his unjustified reputation for being a 'hot Darwinian', there was the additional comfort that all this prosperity had been divinely ordained. The reality that lay behind this bounty was an unprecedented period of expansion in the fishing industry that began within a few short years of Waterloo and was still in progress when the tragedy of 1914 called a temporary halt.

Back in the 1820s, the industry would still have been recognizable as the same small but valuable adjunct to national wealth that led one King of Scots to describe his kingdom as 'a beggar's mantle fringed with gold'. The industry was at its most active where the presence of safe harbours coincided with markets that were within range of a horse and cart or, in large parts of lowland Scotland, the strong legs of the fishwives who carried their wares in the enormous baskets they bore on their backs. Nowadays we would describe the labour-intensive inshore fisheries of the early nineteenth century as artisanal. Dangerous and offering rather poor rewards to its seagoing participants, the white-fish industry at this time posed little danger to the well-being of the stocks and

*Fishwives at Aberdeen fishmarket in
the late nineteenth century*

attracted the attention of the Government mainly as a potential source of ready-trained seamen for the Royal Navy.

All of this was to change with the coming of the railways. Although cities built on estuaries like London and Edinburgh had long enjoyed the luxury of fresh fish landed at Barking and Portobello, now even Birmingham and the great inland centres of the Industrial Revolution could share in the sea's harvest. As Great Britain was to pioneer the Industrial Revolution, so the British fishing fleet was both to pioneer the fuller exploitation of the narrow seas and to sample the greater riches beyond. The Outer and Inner Silver Pits in the North Sea were richly furnished with bottom-living fishes and had been known to fishermen for many years. The Dogger Bank, its name derived from that of the two-masted Dutch fishing vessels which once frequented it, held out even richer prizes, especially cod. The Thames Estuary was so clean that line-caught cod could be landed alive from well smacks, specialized vessels containing a compartment open to the sea to which the cod were transferred on capture. It was not, however, an immediate transfer. The swim bladders of cod – the fishermen called them 'sounds' – are not open to the foregut as they are in more primitive bony fishes like salmon. When a cod is brought to the surface, the gas in its swim bladder expands grotesquely as the water pressure falls. Affected fish are unable to dive, and die as a result. The fishermen learned how to release the pressure with a silver bodkin before releasing the no doubt much relieved cod into the well.

If the Barking men were first to exploit the potential of the Southern Bight of the North Sea and by mid-century had shifted their base to Great Yarmouth, men from the Devon port of Brixham were already established in Hull, having outdistanced their Barking competitors in the North Sea and having at the same time reached Fleetwood on the Irish Sea. In Scotland, meanwhile, the port of Granton gave its name to a type of trawl which was held open by otter boards instead of a cumbersome beam. From that moment on, a vessel's horsepower rather than the length of its working deck set the upper limit to the size of the net it could tow and therefore the fishing effort it could exert. Around the Scottish coast, increasingly adventurous long-line fishermen supplemented the trawlers' catches by landing fish undamaged by the hurly-burly of the cod end. Many of these boats were owned by their skippers, a tradition which has been followed in Scotland until the present day. Some, the so-called great liners, would venture as far as Lousy and Porcupine Banks, oceanic grounds where attacks on their hooked fish by fiercely parasitic amphipod crustaceans – distant relatives of the toe-biters of the Chess – could reduce them to skeletons in a matter of hours.

If large-scale fisheries for species like cod, haddock, plaice or whiting – so-called demersal fishes which spend much of their lives

C.J. STANILAND

Auctioning fish at Billingsgate in the 1880s

close to the seabed – had to await the railway age, it was another story for herring. For most of the history of fishing off North-West Europe, herring was the most important of the mid-water or pelagic species. What made the herring fisheries different was that, from the late medieval period onwards, methods had been developed in Holland for preserving the catch by salting to reduce the water content of the fish so that, with or without subsequent smoking, herring could be prevented from deterioration more or less indefinitely. A large North Sea industry dominated by Holland grew up around the new process, waxing and waning according to the abundance and distribution of the stocks and the politics of European trade.

That there might be something fundamentally different about herring as a fishery resource was first made clear long before by the experience of the Baltic trading partnership known as the Hanseatic League. This group of North European towns had built much of their prosperity on a Baltic herring stock which disappeared in a few short years after supporting some two hundred years of apparently sustainable fishing. For the first time in recorded European experience, mercantile interests were to suffer because trade in fish is based at the upper end of a complex food web ever at the mercy of changes in marine climate.

The great North Sea herring fisheries grew up during the nineteenth century. It was an industry in which Scottish fishermen played a leading part. They preserved the herring using a variant of the Dutch process known as the 'Scotch Cure'. By no means all of their catch was taken in Scottish waters. As the season progressed, they followed the migrating shoals southwards into the southern North Sea, fishing alongside English and Dutch fishermen. The fishery reached its peak with the advent of that most seaworthy of small fishing vessels, the steam drifter. By the time of the First World War, over a million tons of herring were being taken from the fertile waters of the North Sea by British and foreign vessels. As with the application of steam to trawl-fishing in the seabed, Great Britain had shown the way. Could we still be sure that fishing took only a small proportion of the stocks available to it or that we could afford to ignore the effects of oceanographical change? The time had come to find out, and it was by no means certain that, in seeking answers to these questions, Great Britain was in the leading position earned by its fishing fleet.

———

ASKING THE FISH

———

There is no one time or event which could be said to mark the birth of that branch of applied science called fishery research. Once progress had been made in assessing the weight and value of the catches, so explanations were sought for the changes revealed by these statistics. Could there be such a thing as 'over-fishing', a term first used as early as 1854 by Mr J. Cleghorn of Wick when addressing a meeting of the British Association for the Advancement of Science? What part did oceanographical changes play in alterations in the availability of the stocks, especially sudden ones like those behind the collapse of the Hanseatic herring trade? As the nineteenth century drew to its close the information to answer such questions did not exist, but important steps were being taken to gather it.

Through the work of McIntosh and Dalhousie in Scotland and Frank Buckland and Spencer Walpole in England, the British Government had made a limited commitment to establishing the facts behind the apparently continuing success of the fishing industry. This they had done by direct inquiry of the men involved in the trade and, in the case of McIntosh, by 'asking the fish', working out their life cycles and hence their vulnerability to the latest methods of fishing. Where Great Britain had fallen a little behind was in the setting up of programmes of marine research to establish the physical and biological basis for the complex communities of marine plants and animals of which the fish stocks were merely a part. When, eventually, Britain was in a position to make its own contribution, it was to do so via two quite separate strands of scientific endeavour, strands divided by the Scottish border.

Through my involvement with the Gatty Marine Laboratory, I had had my first vicarious encounter with the Scottish tradition. Now I was to make first-hand contact with a branch of the English one. Fish and shellfish stocks are no respecters of national borders and neither, on the whole, are the fishermen who make their livings from them. That two separate traditions of fishery research should have grown up in Great Britain, and been perpetuated until the present day, is a never-ending source of amazement to our continental opposite numbers. It was certainly not the intention of Lord Dalhousie. The first recommendation of his Royal Commission, when it reported in the spring of 1885, was that 'A central authority should be created to supervise and control the fisheries of Great Britain, if not the United Kingdom [which then included the whole of Ireland] and that a sum of money be granted annually to such [an] authority for the purpose of conducting scientific research and for collecting fishery statistics'. The Lords Commissioners of the Treasury agreed, but the prior existence of the Fishery Board for Scotland posed formidable difficulties.

While agreeing in principle with Dalhousie, the Treasury's short-term response was to make funds available for fishery studies to the Marine Biological Association of the United Kingdom, which had opened a laboratory at Plymouth. A plea for 'harmonious action' between the research activities of the Association and the Fishery Board for Scotland was rebuffed by the Board, and so the wasteful schism first reared its head. But, lest it be thought that its perpetuation was solely the fault of intransigent Scots, another sadly missed opportunity to draw things together arose in August of 1901. Earlier that year, a British delegation had returned from a conference in Kristiania (now Oslo) at which the broad outlines of an international programme of fisheries' investigations had been agreed. They have an emphatically modern ring to them:

1. To obtain an accurate knowledge of the seasonal and periodic changes in the waters of the North Sea, the Baltic Sea, the Norwegian Sea and the south-eastern Barents Sea and of their causes;

2. To determine the amount of variation in the character and abundance of the food supply of the food fishes, whether it be planktonic (in the water column) or benthic (on the seabed);

3. To determine the variation in the abundance and distribution of the food fishes at all stages of their life history; and

4. To determine the extent to which those variations were due, either to natural physical causes acting directly on the fish or indirectly through their food supply, or to the operations of fishing vessels in modifying the conditions of reproduction and growth of the fish.

It was intended that the four categories of work would be pursued from national research vessels and that the programme of research would be supplemented by an internationally agreed system for collecting detailed statistics of catches by species and recording both the date and duration of the hauls and the areas where they were taken. The whole enterprise was to be overseen by a new organization to be called the International Council for the Exploration of the Sea (ICES) and based in Copenhagen.

Although there was strong support within the British Government for these worthy international initiatives, the thought that it might shortly have to give them practical expression caught it somewhat on the back foot. Unlike most of the other nations participating in the deliberations of the International Council, the United Kingdom had no dedicated fishery research steamer other than the Fishery Board for Scotland's yacht, the *Garland*. Unrepresentative of the fishing fleet, restricted to inshore waters and thoroughly unseaworthy, the *Garland* was clearly not the vessel to lead the country's fishery research effort into the twentieth century. It was by no means the only problem. The Boer War had cost many thousands of lives and a great deal of money. It had also soured relations with the Netherlands and Germany, key fellow members of the International Council.

Kicking the problem into the long grass by the time-honoured mechanism of setting up yet another Royal Commission was clearly out of the question. The Government had signed the piece of paper in Kristiania and that was all there was to it. A suitable pause for thought was required, and it was provided by the convening of a Committee on Ichthyological Research, whose terms of reference were to inquire and report 'as to the best means by which the State or Local Authorities can assist scientific research as applied to problems affecting the Fisheries of Great Britain and Ireland; and, in particular, whether the object in view would best be attained by the creation of one central body or Department acting for England, Scotland and Ireland, or by means of separate Departments or Agencies in each of the three countries.'

After examining the practicalities of putting a comprehensive programme of British fisheries research in hand, the Committee made the critical recommendation that, desirable as a central authority would be scientifically, the direction of the work should be under the control of the authorities locally responsible for fishery laws, the collection of statistics and day-to-day contact with the fishing industry. In Scotland, the relevant authority was still the Fishery Board; in Ireland, the Department of Agriculture; and in England and Wales, the Fisheries Department of the Board of Trade.

And so it has largely remained, for better or worse. By the time I arrived at Burnham-on-Crouch, the Fisheries Department had long been part of the Ministry of Agriculture and Fisheries, to which the wartime Ministry of Food had been added to form MAFF. The senior permanent fishery official, Ian Graham, was still called the Fisheries Secretary and enjoyed a degree of independence from his terrestrial colleagues. The scientific effort was run from the Ministry's Lowestoft laboratory by Dr H. A. Cole (even his wife did not use his Christian names, but called him 'Nibs'), the Director of Fisheries Research for England and Wales. With Home Rule, the Republic of Ireland had gone its own way, but Northern Ireland ran a fishery research programme with MAFF help, a formula which was also followed by the administration in the Isle of Man.

This is not the place for a detailed account of the different pathways followed by the Scottish and English traditions of fishery research and how both were disrupted and separately resurrected after each of the world wars. Paradoxically, many of the leading scientists in both organizations tended to be English because, until relatively recently, biological subjects were not taught as distinct disciplines in Scottish schools. Over time, the most important difference in emphasis between the two traditions was in the greater effort made by the English laboratories, all of which included the word 'fisheries' in their title, to gain a quantitative understanding of the direct effects of fishing on what are called the dynamics of fish populations. It was a field of inquiry which was to gain world leadership for the English fish population dynamicists in the years following the Second World War, yet the acorn from which this mighty oak was to grow had been planted between the wars by E. S. Russell, the only Scotsman ever to run the English fishery research programme.

North of the border, the Scottish tradition was centred upon the Marine Laboratory in Aberdeen, its title unconsciously indicative of the greater emphasis it has tended to place on understanding the ecology of the sea in its broadest sense. This is a worthy objective, the importance of which is being increasingly realized as the unprecedented fishing pressure of the twenty-first century is affecting marine food webs to a degree that would have been unimaginable in the nineteenth. However, ensuring the relevance of the ecological endeavour to the real problems of the fisheries is, and remains, a great challenge for the Scottish research directors.

Differences in scientific emphasis between institutions are, in principle, a healthy thing, but over fifty years ago Sir Alister Hardy could see the scientific value of combining the best of both approaches in a single coherent organization, an opinion held by every other independent marine scientist who has since looked at the problem. Furthermore, the costs of pursuing modern marine fishery research programmes are now so high that neither the English nor the Scottish programmes are adequately resourced for the demanding tasks which face them. The logistical case for combining both endeavours to create a single world-class organization, equipped with a properly balanced fleet of research vessels, is now as strong as the scientific one.

OIL AND WATER

Where does petroleum come from? The beds of well-established lakes and ponds are often rather short of the oxygen required to promote the rapid decay of dead plant and animal material, and so freshwater ecosystems are less efficient than their marine equivalents in recycling nutrient salts. Instead, they tend to lock them up in layer upon layer of evil-smelling deoxygenated mud. Similar conditions can sometimes arise in marine basins when the fresh sea water cannot bring in enough dissolved oxygen to balance the amount used up as dead material raining down from the surface decomposes. Over time, substantial quantities of pickled material accumulate, and the effect of many millions of years at elevated temperatures and pressures is to reduce the organic matter in these deposits to petroleum hydrocarbons – liquid fossils sufficiently different from the organisms from which they were derived as to constitute pollutants when accidentally released back into the sea. Such are the unconscious ironies of the natural world.

A Hull trawler in the days of sail

The *Torrey Canyon* disaster had left a number of loose ends and, since I was a member of the pollution group at Burnham, part of my responsibility there was to help try and tie these up before another, similar incident caught Ministers with their pants down. A new generation of super-tankers, more than twice the size of the *Torrey Canyon*, was coming into service and the thought of one of these leviathans breaking up in British waters was not a pleasant prospect. My unrivalled knowledge of crustacean toenails was of limited value in meeting this challenge. Fortunately, there were others, much better qualified, to keep me out of trouble. Foremost among them was the physicist J. Wardley Smith, who had spear-headed the *Torrey Canyon* clean-up and taken most of the flak and not nearly enough of the credit in the public debate that followed.

Left to their own devices, most crude and heavy fuel oils that are spilled far out to sea break up as a result of evaporation, weathering and the natural turbulence of the open ocean. By the time such oil reaches the coast, it is in the form of tar balls, a thorough nuisance but a relatively minor ecological threat. Leaving spilled oil alone has a lot to be said for it, although not in the vicinity of bird colonies or tourist beaches. There was never any shortage of enthusiasts for containing the oil with booms, or floating barriers, and pumping it to a place of safety. It's a lovely idea and works quite well for small spills in harbours, but how would it work in the open sea? We soon had the opportunity to find out.

We had crossed the Solent in a small hovercraft and were now on board the lighter on whose deck the test sections of boom lay. It was a sort of double sausage, the top one to be pumped up with air and the bottom one with sea water. The boom lay in great slack lengths down the full length of the deck like a flensed whale. Above was a tall derrick which began to swing with increasing vigour as the ancient lighter rolled in the seaway. The master of ceremonies, who had both invented the boom and chartered the lighter, became more and more excited. Here surely was the answer to the next *Torrey Canyon*. 'Surround the oil with my patent boom,' he advised, 'and then burn it off with this special floating stove which works rather like a rocket,' but on this occasion we had no oil and, fortunately perhaps, the rocket stove existed only as a drawing back in its inventor's office. Meanwhile the derrick was describing great ellipses in the sky, everyone was hanging on like grim death to whatever solid part of the vessel was closest, and the trial was aborted. It was not a good start and the MC, who had kept his peaked yachting cap on throughout, was not a happy man.

The sequel was more professional, if a little less entertaining. It also came at rather an awkward time. My wife was heavily pregnant with our first child and the trials were planned for a fortnight after the anticipated date of birth. After observing the uselessness of the lighter at first hand, we had made contact with the Admiralty, which had put the Experimental Trials Vessel *Icewhale* at our disposal. ETV *Icewhale* was kept at Portland Naval Base, where she acted as tender to the Admiralty Underwater Weapons Establishment (AUWE). We had been allocated the only free week in *Icewhale*'s busy summer programme and there was no question of changing the dates if the obstetrician's prediction proved faulty. It was, by thirteen days. At that time, the modern absurdity of paternity leave had not even been thought of. My sea kit was packed and, birth or no birth, I would be joining *Icewhale*'s ship's company. My wife did not let me down. With the help of an admirable Persian midwife who made the startling announcement, 'I cut you now,' and with me as an interested observer, my wife gave birth to our son John with hours to spare.

HM Naval Base Portland was, and is, a busy place. Its main function is to work up warships to full operational efficiency before they join the fleet. Portland's capacity to put ship's companies through the hoop is legendary and it is a skill for which other navies are happy to pay handsomely. We arrived to find darkly painted German destroyers and Dutch and British frigates, in their lighter grey, moored in the outer harbour. Among them, flying the flag of her captain, a Rear Admiral who bore the rather grand title of Flag

Officer Royal Yachts (FORY) at her raked foremast, was HM Yacht *Britannia*. It was a bright afternoon and her deep blue hull gleamed as she swung at her mooring. After an encounter with a bewildered Ministry of Defence policeman, we eventually found *Icewhale* alongside in the inner harbour. What a contrast she was to *Britannia*; even the salt-smeared windows of her enclosed bridge struggled to gleam. *Icewhale* was an ugly old workhorse in the faded yellow and black livery of civilian-manned Admiralty vessels, a colour scheme directly inherited from Nelson's navy. But her skipper and ship's company were keen to help. The half-baked schemes of crazy scientists were nothing new to them and they knew that, for all her slab-sided ugliness, *Icewhale* would not capsize when she lifted the heavy boom off her deck and into the 'oggin'.

There was only one disconcerting aspect. A technician from AUWE was permanently seconded to the ship's crew, a quiet soul with a beard not unlike Lenin's. His favourite reading material seemed to be the latest copy of the Soviet magazine *Sputnik*, a publication which modelled itself on *Reader's Digest* and waxed eloquent about the joys of life under the benign leadership of the Politburo. No doubt it was all very innocent, but it did strike me as odd that such material was in circulation on board a vessel whose normal task was to assist in trials which sometimes involved some of the navy's latest and most secret underwater equipment. Had the lessons of the Portland spy ring been taken to heart? I hoped they had, went back on deck and admired our mighty neighbour across the dock, the cruiser HMS *Blake*. She was a strange-looking leviathan. She and her sister ships *Lion* and *Tiger* had been built to carry a new type of quick-firing six-inch gun, mounted as their main armament in two twin turrets, fore and aft. But *Blake* had lost her aft (or 'X') turret, and the space had been used to construct a hangar and flight deck to operate Sea King anti-submarine helicopters. The hybrid result was ugly in the extreme.

Our trials were to take place in Lyme Bay, a longish steam each morning and evening for the ponderous *Icewhale*. Getting there required the old rust bucket to negotiate Portland Bill at the southern tip of the Isle of Portland. It is a place of strong tidal streams, and horrible short seas build up there when wind and tide are in opposite directions. Although there were times when my breakfast seemed keen on another look at the light of day, it never got one. The trials were successful in that we deployed the boom without too much trouble and, when we joined it up to form a circle, the oil stayed in it long enough to kill a poor cormorant that popped up from below. The method clearly had some potential for use on a small scale in calm weather, but by the end of the operation both *Icewhale* and our clothes were reeking of petroleum and we were quite certain that trying to use booms to contain oil leaking from a super-tanker wreck would be a logistical nightmare.

Another technique that was popular with the armchair pundits arose from the apparent success of the French in sinking several thousand tons of *Torrey Canyon* oil using powdered chalk treated with stearate (the active principle of soap) so that it would stick to the oil. As fishery scientists, we did not like the idea one bit because of its potential to foul fishing nets and taint their catches. The fuss refused to die down, however, and we ended up having to try it for ourselves. We set off from Burnham in FRV *Nucella*, a futuristic-looking steel vessel designed for oyster dredging which had developed something of a reputation for going aground. Our destination was the Maplin Sands artillery range off Foulness Island where we planned to sink a small quantity of crude oil with soapy chalk in the French manner.

The Royal Artillery knew we were coming but had not told their colleagues on the range. We were about to release the oil when an odd-looking vehicle came pounding down the beach and straight out into the sea in our direction. It was a DUKW, an American-made amphibious vehicle left over from the Second World War, and contained a red-faced company sergeant major. Through a megaphone he announced that whoever was in charge of what we were doing was under arrest. Arthur Simpson and I both put our hands up, not in surrender but to acknowledge our responsibility for the strange goings-on. We were taken to the guardroom in the DUKW, did our best to explain what we were up to and, after a few telephone calls, were released as harmless eccentrics.

Back aboard *Nucella*, we opened the bags of chalk powder so that we could blanket the oil slick the instant we poured it on to the sea. Over went the oil and the chalk and we waited for the tide to go out so that we could inspect the sunken blanket of oil that the French report had led us to expect on the sand. We found very little and what there was lay scattered about in greasy little patches. Later, in a big trial off the Hook of Holland, we found out why. The oil goes down all right, but then a lot of it pops up again. When we tried trawling over the contaminated seabed, the results were as we feared. We dismissed the technique as both inefficient and a fouler of fishing gear.

The fuss about oil lasted longer than any of us expected and involved journeys to both coasts of the United States to meet oil experts and even to Moscow to attend the World Petroleum Congress. Moscow in the grip of the Cold War was not an attractive place. The smell of high-sulphur petrol hung about the streets and everywhere there were men in uniform. Most appeared to be officers in the unattractive square jackets and over-large caps that were adopted by the Soviet armed forces after Stalin's day. Receptions and concerts in the Kremlin temporarily lifted the spirits but once these were over I felt bound to agree with the *Daily Worker*'s Moscow correspondent that it was 'the most uncivilized city in Europe'.

It was a pleasant contrast to be in Brussels and meet the classical scholar, Alan Davidson, then Head of Chancery at the British delegation to NATO Headquarters. I was at a petroleum shindig, an event organized by the alliance's Committee on Challenges to Modern Society, itself set up to demonstrate that the North Atlantic Treaty embraced more than merely military issues. Davidson was in the process of writing the first of his splendid books on fish and shellfish cookery, *Mediterranean Seafood*. He had heard from the Foreign Office that a shellfish expert was due to put in an appearance and was not, of course, to know how limited that expertise really was. He invited me out to lunch and showed me part of the typescript of his book. It was completely different from any previous fish cookery book. In the same volume, authentic recipes based on Davidson's acquaintance with leading Mediterranean fish cooks were combined with scholarly accounts of the natural history of the fish and shellfish themselves. This was not all. There was also a great deal of fascinating historical material, beginning with the classical period and extending into the heyday of descriptive fishery science.

As the meal progressed, we sorted through the various sheets with their line drawings and descriptions of shellfish. I remember thinking at the time that the world was not yet ready for a book of this sort but, not wishing to be discourteous to such an obvious enthusiast who had, after all, bought me rather a good lunch, I agreed to take the sheets home and, if I noticed any little hiccups, to get back in touch. I saw none, but to make absolutely sure, I took a colleague at Burnham into my confidence and checked the sheets with him. He saw nothing wrong with them either, but I thought it would be fun to make a tiny addition and took the liberty of suggesting to the author that he record the fact that the best cockles in the world are Stewkey blues, the Norfolk cockles of my childhood. Not only did Alan Davidson include the anecdote, he also gave me a gracious and entirely undeserved acknowledgement. And how wrong I was about the book. It was to become a bestseller and a culinary classic.

Despite these diversions, pleasant or otherwise, we none the less drew up an oil-spill contingency plan for England and Wales that, for all I know, is still in use. Fortunately, I never had to put it to the test because, in the best tradition of the British Civil Service, I was to be moved to another post before another giant tanker hit the rocks.

Subsequent experience with tanker wrecks showed that there was no one way to address the problems they caused, and that the main victims were nearly always sea birds. Dispersing the oil at sea to help protect these innocents is often bedevilled by logistical difficulties, and the aggressive use of solvent emulsifiers onshore has never been repeated since the outcry that followed the *Torrey Canyon* incident. If there is a consensus, it is that the best way to deal with spilled oil is to recover as much as possible mechanically, and then to be patient enough to allow weathering and the activities of hydrocarbon-oxidizing bacteria to blunt its capacity to do harm.

BACK ON THE FORESHORE

The Burnham-on-Crouch of the 1960s was a rural backwater. Its oyster industry was not what it used to be, despite the best efforts of the Burnham laboratory, but oyster culture was still important and there were still old boys about, one the former first coxswain of an airship, who could remember the days when it was big business. Yachting then, as now, was the largest single employer, both directly on the boats and at the clubs, and indirectly through the yacht builders and chandlers who serviced it. The local families – they called themselves Burnham natives after the oysters, and to distinguish themselves from the increasing number of settlers from London – had lived in the area for generations. The natives were welcoming and friendly, one so friendly to his grubbily attractive daughter that he had had two children by her. In the past there had been a strong tradition of wildfowling in the area. Young gulls in their speckled brown plumage were still known in Burnham as Foulness pheasants in recognition of the days when such 'fowl were sometimes accidentally shot by professional punt gunners and sold on the quiet to butchers who put their minced flesh into sausages.

At the time I was there, the best of the coastal wildfowling was a little to the north on the estuary of the River Blackwater. It was a place that had figured in the writings of the Fenland author James

Wentworth Day. 'Jimmy' Wentworth Day was as politically incorrect as it was possible to be long before the term was even invented. His harmless eccentricities included rounding up a visiting Greek shooting party as potential EOKA terrorists and setting mantraps below the ground-floor windows of his house to catch potential burglars. He wrote entertainingly of wildfowling in Essex, even of an occasion when he accidentally peppered some West Mersea 'fowlers with his eight-bore, Roaring Emma.

I remembered that I still had some home-loaded cartridges for my own ten-bore, although by this stage they were beginning to look a bit motheaten as successive reloadings and the high temperature of the black powder had taken their toll. But it was to be some time before the opportunity arose to give these old cartridges a final day of glory. A first-generation Burnham native, Owen Pugh, had heard that I was a keen 'fowler and smoothed my entry into the local wildfowling club. He himself shot with an enormously long single-barrelled eight-bore of the type they used to call marsh rails. It was a splendid old cannon of which I was highly envious.

Like Owen's marsh rail, my car was also a very old lady. It was an upright Ford Anglia made in 1941, the same model that Adrian Horridge had driven years before in Fife. Most of it worked, including the semaphore direction indicators. One thing that did not was the speed indicator and another was the starter motor. A good swing with the starting handle was normally enough to get old Blackie going, but try it half-heartedly and it nearly tore one's thumb off. One calm November morning we set off for the Blackwater Estuary with our guns laid on the back seat and our boots, still muddy from an earlier foray, on the floor. It was not an especially cold morning but cold enough in the car, which had no heating of any kind. A thin drizzle started which the wipers did their best to clear from the windscreen. They were driven directly off the engine. At low revs, they waved desultorily and even light rain would easily defeat them in the struggle to provide a view forward. At high revs, their actions were so manic that they threatened to fly off their mountings at any moment.

By the time we arrive on the salt marsh, the rain has stopped and the sky is just getting light. We can hear widgeon far out on the mud but forbear to go after them, preferring to take our chance in the bend of a creek which cuts a deep gash in the salt marsh. There is not much water in the bottom of the creek but what there is has concentrated the shore crabs whose scuttlings, along with the sucking noises of the ebb tide, keep us company. Ka-boom! A truly colossal explosion far out in the estuary temporarily silences the widgeon but gives a nearby redshank the fright of its life and it takes off, shrieking its alarm call down the creek. One of the last of the old puntsmen has chanced a shot at the widgeon. Thank God we did not go out on the mud ourselves. At best, we would have spoiled his sport; at worst, we might have fallen foul of some of the heavy shot as it coursed down the estuary like a swarm of angry bees.

Wildfowling is one of those sports where having eyes in the back of one's head would be a distinct advantage. 'Whee-o, whee-o' from close behind, and I turn to see a widgeon drake coming straight for me. There is no time to poke and miss and, as the black powder smoke swirls, the widgeon falls dead on to the remains of the summer's sea lavender.

That dawn flight on the Blackwater was to be the most memorable of our forays on the Essex salt marshes. Our bag, as usual, was tiny but we had heard a mighty punt gun in action. We had also seen the wavering lines of brent geese, members of the dark-bellied race. Quick of wing-beat and no bigger than mallard, their numbers were building up well after a population crash believed to have been caused by a disease of one of their food plants, the beautiful ribbon-like sea grass, *Zostera marina* L. Brent have been protected since 1954 in Britain and since 1966 in France, where it has been estimated that some 2,000 are nevertheless shot every winter. Our own 'fowlers were sorry to lose their traditional quarry. A few diehard Burnham natives were said, for a time, to shoot one brent a year to assert their almost Appalachian sense of local identity and in commemoration of past glories. After a slow start, the recovery of the dark-bellied brent goose population has proceeded so well that the 16,000 of 1954 is now around a quarter of a million. The recovery has largely been fuelled by the grazing opportunities provided by pasture and arable land, and for some years there has been a strong case for returning brent to the quarry list.

If fishing has been called the contemplative man's recreation, so also is the sport of wildfowling. Listening to the cockles spitting in the making tide and watching the distant flocks of waders twisting and turning together like smoke caught on the wind, the mind is liable to wander. Why is it that we can no longer raise the gun to redshank or curlew or down the cormorant speeding low over the sea, for which we used to get seven and sixpence and raw material for a nourishing stew? In debating the terms of the Bird Protection Act of 1954, Lady Tweedsmuir had said that she did not begrudge the 'fowler his curlew, provided she did not have to share his dinner. Would that later legislators had taken the same enlightened view. Neither curlew nor redshank are at all uncommon and both

*The common curlew, the 'sea pheasant' of
the old Norfolk 'fowlers*

are excellent fare, low in fat and with subtle flavours not unlike those of snipe and woodcock. Perhaps the law-makers were influenced by these birds' superficial resemblance to other, less numerous waders like whimbrel and greenshank. If they were, it seems only to demonstrate their ignorance of the high level of ornithological responsibility which is the pride of the modern wildfowler.

As to the cormorant, at one time more highly prized than the greylag goose by the crofters of the Western Isles, the reasons for its protection are equally obscure. Some have suggested that it was a *quid pro quo*, agreed by a British delegation to Brussels, in exchange for a reduction in the annual cull of migrating songbirds running the gauntlet of southern Europe. Whatever the reason, it was a thoroughly bad decision for which the inland fisheries of Britain are now paying the penalty as the cormorant populations explode. Even noble Loch Leven, home to the depleted remnant of the most famous brown trout population in the world, is now host to so many cormorants that it is a candidate for European special protection status for this most voracious of piscivores. Thank goodness that brent geese eat plants not fish.

For Burnham native and Norfolk marshman alike, eel-fishing was one of the ways that the old market gunners kept body and soul together during the long summer months. Yellow eels were their usual quarry, caught by fyke net. It is an ancient way of fishing, using one or more baited cod-ends provided with a fence-like leader of netting to guide the eels to their doom. But then, eels are ancient creatures too.

I saw my first eel, *Anguilla anguilla* L., in a white china bowl on the nature table of a school class reserved for boys who were allegedly less academically inclined than my classmates. Lying quietly bewildered at the bottom of the bowl, as befitted a largely nocturnal fish exposed to the full light of day, it had done its best to appear pale by contracting the melanophores (specialized cells containing black pigment) in its skin. What a temptation it would have been, had a heron been standing where I was. It was a very small eel, nine inches or so in length, and I was not to see it again. By the following morning, the white bowl was empty. Goaded into activity by the falling light levels of the short summer night, the eel had no doubt slithered out of the bowl and on to the wooden floor, only to meet certain death from desiccation between the floorboards.

Neither then, nor at any time since, have I suffered from that abhorrence of the long and slithery which affects so many people that it is believed to be a relic of the fear of snakes, hard-wired into the brains of some of our closer primate relatives. I never actively sought eels out but, every so often as I fished in the dykes of north

Norfolk or in the Grand Union Canal, the large and active fish of boyhood dreams would turn out to be just another eel.

Difficult to unhook from the horribly tangled cast, the eels were even more difficult to kill. Sever the head from the body and, so tenacious of life is *A. anguilla*, that, on a cool, damp day, both ends will continue an independent existence for many hours. Such grisly behaviour is possible because eels breathe as much through their skins as their minute gills and because their tissues are especially frugal in their use of oxygen. How, then, to kill an eel humanely? The secret is to put it in a plastic bin bag and transfer it to the freezer. As the temperature falls, so the eel becomes comatose and dies. Leave it to freeze and, on thawing, its mucus loses its sticky sliminess and can be removed completely by one stroke of a kitchen wipe.

The yellow eels caught by fyke net are eels which are still growing and have yet to feel the earliest stirrings of the hormonal upheaval which will one day transform them into the large-eyed silver eels that migrate to sea in autumn. Viewed from above, the dark olive back of a yellow eel is difficult to see against a muddy bottom. Viewed from below, the reflective light yellow cast of its underside does not stand out from the surrounding water. In contrast, silver eels are liveried for life in the ocean, jet black above and pearly white below, their tissues packed with fat to sustain them in their great migration from a shallow East Anglian Broad to their breeding ground 400 metres down in the seaweed-strewn Sargasso Sea. How they get there, no one knows. Their large eyes and colour scheme suggest that they swim westwards too deeply to make much use of celestial aids, so it is possible that they depend for their navigation on sensitivity to magnetic ones.

What happens when they and their much smaller male consorts – usually too small to be of commercial value – keep their gloomy tryst is as mysterious as the coupling of Barbara Cartland's heroines. Unlike the latter, though, once their passions are spent, every one of the poor eels dies. Despite the best efforts of many an expensive research cruise, no one has seen the fertilised eggs or newly hatched larvae, called leptocephali, among the tangle of seaweed, but since the great Danish oceanographer Johannes Schmidt first captured the leaf-like leptocephali outside the Mediterranean, he and his successors have charted their return north-easterly drift across the Atlantic to found another generation of eels in the fresh waters of Europe. Probably only a few millimetres long at birth, by the time the larva drifts as far as the edge of the European continental shelf, a journey which takes at least a year, it may be as long as eight centimetres.

An early representation of the leptocephalus larva of an eel

It was just such a one that lay atop the catch of our plankton net on a brief night in June 1996. FRS *Scotia* wallowed whale-like in the long Atlantic swell as, lovingly, we lifted the dying larva from among the pulsing medusae (small jellyfish) and drank in its beauty in the yellowish light of the ship's laboratory. Had it lived, the larva would have metamorphosed into a transparent glass eel to develop pigmentation and enter a Scottish or Norwegian river as an elver the following spring. But Scotland and Norway lay way to the east that night and the continental slope over 1,000 metres below our rolling keel.

The leptocephalus was shaped like one of the laurel leaves Peter and I used to crush to kill trapped butterflies. It was completely transparent apart from its eye, a tiny jet-black pupil rimmed with an iris of brightest silver. Under the microscope we made out a surprisingly ferocious-looking set of needle-like teeth. We also discovered the secret of the silver iris. As with the first trout of my childhood, rows of guanine crystals provided the gleam. Lit from above, they would act as mirrors and thereby make the large eye, apart from the small pupil, appear as transparent as the rest of the leptocephalus. We could tell from the orientation of the rows of crystals that, for the mirror effect to work, the larva would have to swim at an angle of some 30 degrees to the horizontal. It was a new observation and, with *Scotia* costing thousands of pounds a day to keep at sea, rather an expensive one. Mother Nature has ever been a coy mistress and to see her naked you have to pay the price.

———

THE SNOWFIELDS
OF CORNWALL

———

Tanker accidents sell newspapers and kill a lot of sea birds but, for most marine organisms, they are not the most important form of pollution or even the most important form of pollution by oil. Once the public hysteria over the *Torrey Canyon* had died down, it was time to turn to other environmental issues potentially more threatening to the well-being of fisheries.

The bed of the shallow seas which swirl around our islands is home to the demersal fishes – species like cod and haddock and the many sorts of flatfish which depend for their food on the worms and shellfish that live on or in sediments. Just as the productivity of farmlands ashore depends upon the type of soil, so the richness of the seabed as a feeding ground for demersal fishes depends upon the type of sediment. Muddy sands support the richest feeding. They are seen at their best in places like the southern North Sea where the underlying geology is sedimentary and naturally includes a fair proportion of clay relatively undisturbed by wave action. The western English Channel is a very different place, especially off the Cornish coast. Rocky outcrops abound and, between them, the softer sediments are regularly winnowed and coarsened by the grounding of waves large enough to attract surfers. There are better places to be if you are a flatfish that makes its living by nibbling the siphons of mud-loving molluscs. We were shortly to encounter the exception that proves the rule.

The landscape around St Austell and Mevagissey bays in Cornwall is one of the most extraordinary in England. Even in the hottest of summers a range of snowy peaks, as white as the Cairngorms in winter, extends down to the sea. The oddest feature is that the 'snow' continues across the full width of the beach, impervious to the action of the roughest sea, which is itself unusually pale. The snow is china clay waste and we were about to take a hard look at its effects on the seabed.

MFV *Christabel*, the vessel we had chartered for our survey, was typical of many of the smaller fishing boats working off the Cornish coast thirty or so years ago. She was about forty feet long, had her wheel-house aft and a Gardner diesel engine that had run for over twelve years without overhaul or, indeed, the slightest

The snowfields of Cornwall

indication that it required one. Her home port was Mevagissey and she normally made her money potting for crabs and line-fishing for mackerel. Like most of the boats down that coast, she had a square transom stern, not because it was any better or worse than the cruiser stern favoured by most of the small fishing vessels off the Scottish coast, but because it had been the fashion for as long as anyone could remember.

It was mid-afternoon before we arrived. The formalin preservative was in large polythene jerrycans which we were glad to unload from our Land-Rover: the top of one had started to seep and the horrible acrid fumes were making our eyes water. *Christabel's* skipper, Joe Chesterfield, was pleased to see us. He had had a long association with the Burnham laboratory, for which he provided details, not just of his crab catches, but of the effort which went into obtaining them. Joe's statistics were combined with those of other skippers to give an index of the abundance and size structure of the crab stocks of the western Channel. Surveying the seabed was something new for Joe but nevertheless welcome – it paid a little better than crabbing, there was no precious time spent trying to get hold of suitable bait and no dealer to hold him to ransom.

The Baird grab has long since been superseded by supposedly better devices for sampling the surface of the seabed. Its great advantage over its successors was that its jaws moved sideways on rails instead of moving in an arc like those of most other grabs. To look down on a Baird sample was to look down on to an undisturbed piece of the sea bottom. The snag was that the top was open and, theoretically, anything big and mobile that did not want to join us aboard *Christabel* could, like the Sunday tabloid reporter, 'make its excuses and leave'. Joe, though, was an artist who was as keen to see what lay under our keel as we were. The grab was brought gently to the surface as though every haul might have contained King John's treasure.

What we saw surprised us. It had looked as though nothing could live in the blanket of white that spread down from the Cornish Cairngorms, across Pentewan beach and out into the bays of St Austell and Mevagissey. Not much did live close inshore, but farther offshore, where the rain of white silt was manageable for those marine animals adapted for life in muddy sediments, a rich community had established itself. It included a wealth of marine worms and small molluscs and large numbers of the burrowing sea urchin, *Echinocardium cordatum* (Pennant), and the beautiful brittle starfish, *Amphiura filiformis* (Müller). Back in 1918, the marine biologist C. G. J. Petersen had recognized this particular association of animals as characteristic of some of the richest feeding grounds for flatfish in the North Sea. The trawlermen also knew about burrowing sea urchins and had learned to associate their presence in the cod end with good catches. Fishermen and scientists speaking with one voice? Only up to a point; for the toilers of the sea, the scientists' *Echinocardium cordatum* (Pennant) had always been 'Dutchmen's farts'.

That pollution by spilled oil is wholly bad no one doubts, and there are still many environmental purists who deplore any industrial or agricultural process, the disposal of whose wastes alters what was there before. What we had seen in Cornwall was an instance in which the effects of discharging an unsightly by-product had unexpectedly turned a small corner of the English Channel into something resembling one of the more fish-friendly banks of the North Sea. However, so small was the corner that it would not have supported the activities of a single modern trawler for more than a day or two. If the effects of china clay waste are not entirely bad, neither are they convincingly good enough to balance their horrible effect on the natural appearance of England's wildest coastline. One up to the purists.

RED, WHITE AND BLUE

One of the problems of using Burnham-on-Crouch as a base for maritime operations which could be anywhere around the coast of England and Wales, was that everywhere else was a very long way away. We had been asked to look at another messy waste-disposal problem off the north-east coast of England. Our heavier gear had been sent on ahead, so we were spared the four-wheeled kettle-drum of our customary Land-Rover and motored north in the under-powered comfort of a Morris Traveller. We arrived at

Hartlepool in the afternoon and searched the docks for the fifty-five-foot Fishery Research Vessel *Tellina*. We found her at the bottom of an iron ladder, slippery with green slime. Her colour scheme of buff upper works, black hull with a white line and deep red bulwarks was the same as that of *Icewhale*, the red bulwarks an unconscious echo of the sailing navy where gun decks were painted red so that the sight of blood did not depress the gun crews during an action.

The wooden Fishery Research Vessel Tellina

Like *Icewhale*, *Tellina* also flew the blue ensign but in *Tellina*'s case the 'fly' was defaced by the badge of the Ministry of Agriculture, Fisheries and Food. (The details of the badge, which featured a codfish, were not easy to make out because of soot from the diesel exhaust.) The blue ensign is flown by all government vessels other than ships of the Royal Navy, which fly the white ensign, a privilege also accorded to members of the Royal Yacht Squadron based at Cowes. All other British vessels have the right to fly the red ensign. In Nelson's time, when *Tellina*'s livery would have been quite the latest fashion, the navy's ships flew blue, white and red ensigns depending upon the seniority of the admiral in command. The blue ensign was reserved for junior members of each flag rank. Newly promoted Rear, Vice and full admirals all flew the blue ensign. With seniority came the right to fly the white and then the red ensign. (Nelson was a Rear Admiral of the Blue at the Battle of the Nile – where, confusingly, he flew the white ensign because he thought the blue one would not show up in the failing light – and a Vice Admiral of the White at Trafalgar.) A full Admiral of the Red was accorded the title Admiral of the Fleet, a designation which was later formalized as the ultimate flag rank. It is no longer conferred and will die out with the holders, who currently include HRH Prince Philip and once included Kaiser Wilhelm II.

Viewed from the top of the boarding ladder, *Tellina* did not look too bad. Admittedly the stern looked as though it had been intended for a larger boat and the wheel-house, which was long and low, had the dated, streamlined look of a 1930s caravan, but otherwise it was not immediately obvious that there was anything radically wrong with her design. I was met, at the bottom of the ladder, by the cook, Jack Reynolds. He had the worst job on board. His tiny galley was stiflingly hot – bad enough on a winter's afternoon like this but a truly terrible place on a still day in mid-summer. It was his task to produce three hot meals a day on a frugal budget, supplemented only by any fish or shellfish taken during research cruises. I should, perhaps, explain what is meant by the term 'cruise' in this context. For a start, it has absolutely nothing to do with touring the fleshpots of the Mediterranean in a luxury liner. When a vessel sails from its home port to another, it has undertaken a 'passage'. If it then returns home, the two or more passages constitute a 'voyage'. A 'cruise' is when a vessel leaves a port, goes out to sea for a bit and then returns to the same port. That's what we were about to do. Interestingly, as the man holding the baby, so to speak, I was accorded the title of Naturalist-in-Charge, a designation that would have delighted McIntosh.

As Jack showed us over *Tellina*, it became clear that the oversized stern section was to provide sufficient accommodation for seven people: the skipper, engineer, two deck-hands, a cook and two scientists. The skipper and crew all slept in bunks arranged in the side of the boat around the small table where we ate our three meals and enjoyed our invariable nightcap of strong cocoa and a piece of yellow cheddar cheese. The cocoa was a reminder of the crew's service in the navy: hot kye made of grated, unsweetened chocolate was a real life-saver when watch-keeping on an open bridge. As scientists, we had the privilege of minute cabins to ourselves and, between our accommodation and the crew's, there was a little space which contained a small, fully enclosed bookshelf with about half a dozen hardback books, all of an improving character. On a cream plastic panel screwed to the bookcase were the words: 'Donated by the Girls of Bedford School'. I rather suspected the choice of books was not that of the girls but of some frumpy old schoolmistress.

Next to come aboard was Jim Taylor, a dark-haired Scot from Stonehaven in the Mearns. Jim was the engineer, responsible for the main and auxiliary engines and all the other mechanical and electrical equipment on board, and a meticulous one: never, in my experience, did anything of Jim's break down. As well as being an engaging companion ashore, where he enjoyed long walks, he had a remarkably retentive memory. His normal greeting would be to recall the last occasion he was in your company and what you were eating at the time. The two deck-hands, Chas Mullender and Charlie Button, arrived not long after Jim. Chas was well into his sixties and, like all of the crew except Jim, a native of the Suffolk–Norfolk border. In character he was not unlike Private Godfrey of *Dad's Army*, a gentle soul who hated anything in shipboard conversation that was in the least smutty. He was, for instance, greatly scandalized when Charlie inadvertently characterized one of the participants in the then topical Profumo scandal as being 'hot round the bum'.

Charlie Button was from Hulver, a tiny agricultural village in Suffolk, and before the war had owned his own boat. He had fished the eastern Channel with it, keeping a careful record of sediment types and planning his tows accordingly. Like all of *Tellina's* crew, Charlie had served during the war in 'Harry Tate's Navy', as the Royal Naval Patrol Service was popularly known; it was an organization largely composed of fishermen and specialized in mine-sweeping and escorting convoys in the narrow seas. Although diminutive, Charlie was absolutely fearless. Exceptionally, as a petty officer, Charlie had had his own command and dealt successfully with a German Junkers Ju88 bomber which was unwise enough to take him on.

Last to come aboard was the skipper himself, Billy Burroughs. Billy had been sent out by Jack to get some groceries which he brought back in an old black oilcloth bag. He was wearing a long mackintosh coat and had a hand-rolled cigarette on his lower lip. In appearance Billy was not exactly Admiral Sir David Beatty – the famously sartorial commander of the British battlecruisers at Jutland in 1916, who wore his richly gold-braided uniform cap at a jaunty angle and six buttons instead of four on his specially tailored monkey jacket – but he was highly sensitive to the needs of marine scientists and an expert at converting their ill-expressed ideas into well-planned action. After our evening meal, Billy and I spread out a couple of Admiralty charts in the wheel-house. I explained that north of the Tyne, near to the little town of Blyth, large quantities of a fine grey ash, produced as a waste product by power stations, was dumped directly into the sea. No one knew what effect the dumping of this fly ash had on fisheries, only that the Minister had been asked about it and wanted it investigated. While we were at it, we had also been asked to take a look at the dumping of coal waste over the cliffs and on to the beaches between Hartlepool and Seaham. It would be a cruise, all right, but not one likely to attract many paying passengers.

We decided to undertake the fly ash survey first and sailed for the Tyne Estuary, which was to be our base, early the following morning. On our way north, we passed the cliffs over which the coal waste was being dumped and went as close inshore as we dared. In a scene straight out of the eighteenth-century dawn of the Industrial Revolution, hopper after hopper was pulled along on a conveyor belt and emptied its contents on to the beach with a great clatter. The sea around was a sooty grey, and we resolved to return and find out what, if anything, lay underneath it once we had completed our work off Blyth. We secured alongside at South Shields, drank our strong cocoa and I turned in without a care in the world.

I had not listened to the shipping forecast so got something of a shock the next day when, as we passed through the breakwaters that guard the Tyne, *Tellina* was picked up as if by a giant fist and smashed down again into the trough of a big sea. Billy looked at me, I thought a little anxiously, but in my ignorance I indicated that we should stick to the plan we had roughed out back in Hartlepool and continue to the ash-dumping area and take samples. Once we were clear of the Tyne, the boat settled down, taking some water over the bow but able to maintain eight knots or so. We started the survey. *Tellina* by this time was wallowing in a peculiarly nauseating manner. I was sick, my scientific colleague was sick and even Chas Mullender, the oldest and most experienced man aboard, was sick. We nevertheless completed our transect of stations, sampling the seabed at each with a 300-pound grab which swung alarmingly as it came inboard. Our line of stations went right through the dump site and we returned to the Tyne with the tide mercifully behind us.

That evening nobody said very much, but when Billy asked what we should do the following morning, I submitted that we should work another transect of stations through the dump site but at right angles to the one we had just completed. We sailed before breakfast, the passage through the breakwaters even more violent than that of the morning before. We completed the transect, again involuntarily fed the fishes and, after a dreadful passage, during which *Tellina* showed every sign of wanting to take us on a visit to sample the seabed at first hand, secured alongside. Unknown to me, we had put to sea in a vessel with inadequate buoyancy forrard and in the face of a storm force ten warning. We had got away with it thanks to the skipper's seamanship and because, as usual, the shipping forecast was a little ahead of the game.

As it was, the weather blew up so badly that *Tellina* was storm-bound in the Tyne for days on end. When the gales eventually blew themselves out, it was too late to have another look at the sea of soot. *Tellina* returned to her home port of Lowestoft and we took our hard-won samples back to the laboratory. In the hurly-burly of their collection, we had had little time to draw conclusions about the effects of the dumped fly ash, but our impression was that its effects were not at all well-defined. The particles were so fine that their sinking rate was low, and the effect of the tide would therefore be to spread them over a wide area. That, along with the fact that the sediments naturally present in the dump zone were also fine, made interpretation much more difficult than in the case of the china clay.

One odd feature of the fly ash, though, was that, in some of the samples taken right in the middle of the dump site, the rain of ash had formed a sort of friable concrete. The concrete preserved a record of the burrows of crustaceans common in the area with the strange name of *Upogebia deltaura* Leach. Just as aircraft dominate the modern battlefield, so large, fast-moving predators like dogfish, cod and seals pose a serious threat to creatures scuttling over the seabed. Burrowing provides some protection from such dangers and many crustaceans have adopted it. None is more successful than *Upogebia*, a distant relative of the hermit crabs which protect their soft abdomens using the shells of dead whelks and sea snails.

As they grow, so hermit crabs need to change their shell for a larger one. For *Upogebia*, a burrow, tailor-made to fit, takes the place of a borrowed shell, the scarcity of which can often spell the end for hermit crabs. The numbers of *Upogebia* are limited by the availability of mud of just the right consistency to support burrows without collapsing. At the dump site, some of their burrows were very contorted, indicating that the patient *Upogebia* repeatedly had to dig themselves back to the surface of the seabed. This was one of the few definite indications we had that the dumping was harming the bottom fauna. It was not enough on its own to put an immediate stop to the dumping of fly ash in the sea, but it was telling evidence that we would not have got without Billy Burrough's stout heart in the face of a weather warning he knew about but I, in my inexperience, did not.

How careless we can be in disposing of waste materials at sea. Our vision stops with sunlight glinting off the surface of the water, but the effects of disposal are far below. Recording them, so that their true costs can be set against the profits from the industrial processes which give rise to them, continues to be a great challenge to marine science and an even greater one to governments.

OF SHIPS AND TROUT

When I first joined *Tellina*, she was a relatively new vessel, built at the Moray coast yard of Jones of Buckie. Unfortunately, the shipyard was not responsible for her design. Her specification required a vessel capable of deploying a wide range of fishing and sampling gears and accommodating the five seamen and two scientists required to work them. All this versatility was to be contained within a hull which was only fifty-five feet between perpendiculars and with a draught shallow enough to enable her to use tidal harbours liable to dry out at low water.

For once, and it was to be the only occasion during his long career, the Ministry did not subject the design team to the wise counsel of 'Gentleman' Geoffrey Trout, a leading scientist at the Lowestoft laboratory who had an unblemished record for overseeing the design of first-class fishery research vessels for use all over the world. Had he been involved, he would undoubtedly have warned against so blatant an attempt to get a quart into a pint pot. As it was, the manager of the shipyard had the thankless task of turning the resulting set of plans into larch and oak. He expressed his misgivings at the launching ceremony in the immortal words, 'I'm surprised the monstrosity floats.' It did but adopted the disturbing head-down attitude of a potential Davy Jones Express. Steps were taken to alter its trim but the basic problem, lack of buoyancy forrard, was never adequately addressed. I have lost count of the number of vessels in which I have sailed, but only *Tellina* carried with her that morale-sapping sense of being 'Nearer my God to Thee' every time the wind got up.

Apart from the ability to stay afloat in all weathers, the most basic requirement for a fishery research vessel is that it should be capable of taking samples of fish representative of those taken by the commercial fleet – and to do so it must use the same fishing gear as the fleet. Rarely is it necessary for research ships to retain large catches for longer than the time required for the scientific staff to sort through them: there is no need for the large storage capacity built into many fishing boats. Theoretically, therefore, the design team for a fishing research vessel can use the space not required for fish-rooms for the laboratories and specialized equipment needed to pursue fishery and oceanographic research programmes without compromising the sea-keeping qualities insisted upon by fishermen since the days of sail.

Until the Salveson family invented the modern stern trawler with a stern ramp, a layout based on that of the whale factory ship, trawlers shot and hauled their trawls over the ship's side. Sufficiently broad in the beam and well-ballasted not to broach-to when presenting its side to wind and waves at these critical times, but streamlined from stem to stern so as to work with the sea on passage and when fishing, only the steam drifter has ever approached the seaworthiness of the traditional side-trawler. The sea-kindliness of the design was seen at its best when a reciprocating steam engine, usually a three-cylinder compound, provided the power. The effect of meeting a big sea was to cause a momentary slowing of the propeller until the wave had passed, foaming harmlessly, down either side of the ship. A motor trawler cannot ride with the punches in this way but tends to press remorselessly on, especially in the insensitive hands of an automatic pilot, in a shuddering welter of spray.

With the side-trawler layout as a starting point and a slow-revving Gardner diesel to provide the power, building seaworthiness into the hapless *Tellina*'s design should have been straightforward enough, even if it was to prove otherwise. But when the starting point is a small steel stern trawler, a far greater challenge is posed. Early versions of such ships tended not to have the fine entry and flared Atlantic bow of their side-trawling predecessors and, whereas the sea would close behind the streamlined counter or cruiser stern of the side-trawler, it often boiled in turbulent fury in the wakes of the broad square sterns of the new ships.

FRV *Corella* was the first British fishery research vessel to follow the new fashion. To the surprise of the conservatives, she was a triumphant example of Geoffrey Trout's ability to convert the ostensibly incompatible demands of scientific prima donnas into a highly effective, seaworthy unit that did everything demanded of it and looked good as well. Marine biologists who think like a naval architect are rare enough. This was part of Geoffrey Trout's genius, but it was not the whole secret of his success. There was something about his demeanour – an air of effortless superiority which was also authoritative – that immediately induced in stroppy scientists that 'dropping-down-deadness of manner' said to be so prized by bishops in their curates. Indeed, so great was this power that it once inadvertently worked its magic on the Hull trawler fleet.

Trout was among the Hull men off Iceland in FRV *Ernest Holt*, a steam Arctic trawler built for research but with the same external appearance and capabilities as the best of the commercial fleet. As usual, the skippers were chattering away to each other on their bridge radio sets.

''Ow are you doin', George? We 'aven't got owt,' said one. At the time sixty baskets of prime cod were on their way down to the fish-room.

'We've got fook all, Jim. I reckon them fookin' yellow bellies [Grimsby men whose nickname was derived from the yellow facings on the uniforms of the Lincolnshire Yeomanry of Victoria's reign] 'ave cleaned fookin' place out.' George's fish-room was filling up nicely by this stage of the trip and not just with cod. He had a fair pile of jumbo haddock as well.

'We live in hope, we live in hope,' offered another perverter of the truth. His fish-room was almost full and another day would see him ready to dash home to catch the best of the market.

'Would any of you fellows care to give me details of your catches?' asked a voice so plummy it would not have been out of place in the precincts of Canterbury.

The air went silent for a full two minutes. Eventually George could stand it no longer. 'Fookin' 'ell, oo's that?'

It was, of course, 'Gentleman' Geoffrey and such was his charm that, before they knew it, the skippers had given him all the information they had been keeping from each other for the best part of three weeks.

Corella was 117 feet of welded steel, built as a stern trawler around a diesel engine of 1,000 horsepower. She was a big ship in miniature. Her superstructure was so arranged that, although the bridge structure was well forrard, it still had a good view of the working deck aft with its winches, derricks and stern ramp for the trawl. It was a fine design, marred only by a large square gantry on the stern – a useful device in itself, but this one was on the big side (the rumour was that it had originally been intended for another ship) and some blamed it for *Corella*'s only vice: a tendency for her short, squat hull to pitch violently in a head sea. It was to be some time before I was to step aboard this splendid little ship. In the meantime, it was back to the Davy Jones Express.

———

THE SEA TOMATOES

———

The long sandbanks which spread out, like underwater fingers, into the Southern Bight of the North Sea guard the Outer Thames Estuary. It is an area of strong tides and the short, violent seas that rapidly build up with the gales of winter. For animals living on the bottom, it is no place to settle permanently. Forms like catworms, swimming crabs and brown shrimps – creatures that live on and in the surface of the sand – do best, making a tenuous living at slack tide and burying themselves at other times to avoid the turbulence swirling around them.

Ever since 1887, sewage sludge from London had been dumped in the centre of this area, in the Black and Barrow Deeps. By the early 1970s, some five million tons of the stuff were being dumped there every year. A similar amount of this unattractive material was being dumped in deeper water off New York. My friend, the American marine biologist Jack Pearce, had examined the seabed there. His findings were disturbing. The area at the centre of the dump site was blackened and 'lifeless' – except for what are called anaerobic bacteria and a few hardy specialized worms which can manage on a minimum of dissolved oxygen. Perhaps similar conditions prevailed in the Barrow Deep. It was to be our job to find out.

I had already seen the dumping operation at first hand aboard one of the ships whose responsibility the task was. I had expected a rust-streaked hulk crewed by East End cutthroats only too familiar with rat-catcher's yellows and, no doubt, many of the other vile infections to which I imagined their calling exposed them. I arrived at the dockside to see it occupied by a single gleaming vessel. It had the fine lines and the understated elegance, on a larger scale, of a Trinity House tender. Where on earth was the sludge boat? I had come to the right place and for once my watch, which had been immersed in the sea on several occasions, had not let me down. I walked up and down the dock getting increasingly anxious. 'Are you the scientist?' I was being hailed from the gleaming vision by a Merchant Navy officer in his smart summer rig. 'Yes,' I yelled back, 'I'm waiting to join the sludge boat.' His reply astonished me, 'This is it, welcome aboard.'

I was taken to meet the Master on the wide, enclosed bridge. It was exceptionally well-equipped, as it had to be to operate safely in what were still, at that time, very busy sea lanes. Back in the saloon,

I was sat at a long table covered with a spotless cloth. A white-uniformed steward appeared, his buttons agleam, and presented a printed menu. By this stage I was beginning to feel distinctly scruffy in my fisherman's jersey of speckled grey – a feeling reinforced when, after we were safely clear of the dock, some of the officers joined me in the saloon. After an impressive three-course luncheon, I went on deck. The ship was rock-steady as we gathered speed in the tideway and remained so as the Barrow Deep approached. Our wake changed from bubbling white to slate grey as the sludge was discreetly discharged and, going smartly about, we made our stately way back to the dock. On the face of it, it looked as though sludge-dumping was far from the disgusting process I had imagined, but would the creatures on the seabed agree?

The next time I saw the elegant form of the sludge-dumper was from *Tellina*'s pitching deck. Chas Mullender had retired, wishing for 'a few years at home with the wife', and been replaced by Jack Hood, another Harry Tate's veteran and a fine seaman and shipmate. Poor Chas's wish was not long granted, for he soon slipped quietly away to sail some celestial sea. At least, he would no longer be bothered with seasickness. But not everyone aboard was pleased that the cruise was to be spent in 'mud hunting'. Jack Reynolds, the cook, hated the mess that came down the companion-way into the accommodation, and the constant stopping and starting required to take sixty-five grab samples was a strain on everyone's nerves. The worst of it was that all there seemed to be to show for it was a couple of boxes of Kilner jars, and these would not reveal their secrets until we were back at the laboratory.

The beam trawling was more fun. There were no boring sediments to sort through and in one haul we caught a large common sole, *Solea solea* (L.), the sublime Dover sole of the fish trade. 'You don't want to eat them buggers. They'll make yer deaf, bor,' someone commented when this aristocrat was spotted. 'They ain't n'good,' said another, 'they always eat tough.' It was nonsense, of course, but legitimate gamesmanship from those anxious to add the doormat-sized sole to the fry they were going to take back to Lowestoft.

Partway through the cruise, we left the Outer Thames with its clanking buoys and flimsy-legged forts (one subsequently became the pirate radio station, Radio King) and made our way up the Orwell to refuel at Ipswich. The May blossom was in full bloom, mostly white but a flush of pink here and there. Its nutty scent followed us past Pin Mill, with its Dutch landscape of sailing barges, and almost the whole way to Ipswich. We spent the night alongside. There was time for a short run ashore, but we were all tired and glad of the cocoa and cheese that ever marked the close of the day. Charlie Button and Jack Hood recalled the great days of

Selling herring at Great Yarmouth in the heyday of the East Anglian fishery

the East Anglian herring fishery. In their eyes, the long-obsolete steam drifter was 'the best sea boat ever built' and drift-netting 'the fairest method of fishing there is'. Someone else set off on a different tack, chiming in with 'Poor old . . . , his old gel, she don't believe in it.' It was time to turn in.

Charlie, Jack and the others were children of the steam-drifter generation, contemporaries of the new ships fishing in the years before the First World War. It was a time of great prosperity in the herring fisheries. The new vessels used the same method of fishing as their sailing predecessors but they could handle greater lengths of drift net and, although limited by wind strength, were not slaves to its direction. The result was a great increase in the herring catch at a time when the East European market for cured herring was still buoyant. Just how long the stocks could have stood up to this increased fishing intensity was never really tested. From 1914 until the end of the First World War, more and more of the fishers of herring became fishers of mines – and any chance of renewing their pre-war prosperity after the Armistice was ruined by the Russian Revolution of 1917, which crippled the market in cured herring.

There was a reduced but still substantial herring fishery between the wars and all of the *Tellina* men took part in it. For the average deck-hand, though, it was a poorly rewarded affair and Charlie left it to buy a tiny inshore trawler of his own. Like many a good wooden boat, it had been built in France, and with it Charlie scratched a living among the soles, plaice and dabs off Rye. The Second World War ended his idyll and also that of many of the herring drifters. The fishery restarted once the mines had again been cleared, but new methods of fish detection and a dangerous new practice of trawling the massed shoals over their gravel spawning grounds threatened the herring stocks as never before.

Within a few short years it was all over. The adult herring stocks collapsed in the face of the fishing onslaught, and the larval herring on which future stocks depended suffered from one of those periods of reduced natural survival which are of no consequence when stocks are robust but are critical when they are not. Herring fishing in the North Sea was banned for many years. The stocks eventually recovered but, with both processing plants and home markets for herring seriously affected by the ban, the industry never regained its former prosperity. It was a sad end, made all the sadder by the fact that Sir Alister Hardy, doyen of the herring scientists and trainer of many of them, lived to see it.

The keen air of early morning swept away the bittersweet dreams of drift nets vibrant with herring. Our quarry now was the seabed itself and, from what we could see of it, the Outer Thames Estuary was rather a forgiving sort of place. At a grab station close to the dump site we had picked up a few dead gobies and common starfish in one of the beam trawl hauls. The grab sample also contained a single dead shore crab, *Carcinus maenas* (L.). It looked as though a load of sludge had drifted over the sandbank separating the Barrow from the Black Deep and temporarily deoxygenated a small area on top of the bank. The living community on the bed of the two Deeps looked remarkably normal. There was no apparently lifeless epicentre as there was off New York, and no sign of the capitellid worms which exploit the edges of such areas.

It seemed incredible that so much sewage sludge could be dumped without obvious signs of devastation. The physical measurements we also took tended to confirm the biological ones. We measured high levels of oxygen from the surface to just above the seabed. Only once did we see any evidence of lower levels and that was at the surface, in water discoloured by a very recent dumping.

The twist in the tail came when we looked in more detail at the samples of the seabed. Under the binocular microscope we noticed what looked like tiny grey-green hedgehogs. They turned out to be the weathered husks of tomato pips that had braved many a cockney colon. They were so numerous that we were able to use them to trace the distribution of the sludge. The pips also allowed us to reassure the Port of London Authority that their ships were releasing the sludge where they said they were and not dumping short and going home earlier and richer. So long as the people of London were allowed to use flush toilets, it looked as though there were worse places to dispose of the final result than among the powerful tidal currents of the Outer Thames Estuary.

Somehow, this remarkable area of shallow sea was able to accept five million tons of deoxygenated gruel year after year without suffering the dire ecological consequences which common sense would have predicted. The answer lay in the power and turbulence of the tidal currents which rapidly dispersed the dumped sludge among a much greater volume of clean sea water saturated from surface to bottom with oxygen. That same turbulence also spread the relatively small solid component of the sludge – it is over 90 per cent water – thinly over the bed of the Barrow Deep and the sandbanks which bounded it. But the turbulence had another, subtly protective effect. Sewage sludge contains high levels of nutrient salts like nitrates and phosphates, substances with the potential to stimulate the explosive growth of algae, including forms which contain the dangerous toxins that cause paralytic and amnesic poisoning. However, nutrients cannot trigger algal blooms on their own. The algal cells also need a lot of light and, in the Outer Thames, this is denied them by the high natural levels of entrained silt stirred up from the seabed by the tidal currents and grounding waves.

Studying the fauna of the seabed was all very well and I learned a lot from it. But it does not appeal to the hunting instinct in the way that real fishing does. The exciting part of the *Tellina* cruises was always the trawling, however unpromising the prospects.

A modern trawl net is a much more subtle instrument than it appears. To the layman it just looks like a rather elaborate way of sieving a lot of water and for a long time that is how many of the fishermen themselves thought of it. The first hint that trawls did not catch fish just by sieving them out of the sea, came with the discovery that catches were greatly increased if the otter boards that hold the wings of the net open were separated from the wing ends by a length of cable. Underwater observations were later to show that the otter boards and their cables served to herd the fish into the path of the net, so greatly increasing its catching power. The second surprise was that fish in front of, or inside, the main body of the trawl would keep station on the various ropes and meshes as the gear was towed forward by the ship. What happened next depended upon the speed of the trawl relative to the endurance of the fish.

Most fish can swim at rates of up to three body-lengths a second more or less indefinitely. For this gentle cruising, they use the bands of red-muscle tissue that run just under the skin on each side of the body. This special tissue has a good supply both of oxygenated blood and of the oxygen-carrying pigment myoglobin. As a result, red muscle does not run out of oxygen when working. When fish are required to swim at greater speeds than three body-lengths a second, they have to employ the white-muscle tissue that makes up most of their body mass. At first it works well and, for a time, is capable of propelling many fishes at rates as fast as ten body-lengths a second. However, for all its short-term effectiveness, the white tissue lacks the special adaptations which keep red-muscle tissue oxygenated even when it is active. Thus, fish required by the speed of the towing vessel to keep station on the trawl at speeds greater than three body-lengths a second build up an oxygen debt just like a human sprinter and soon tire. They attempt to escape by abandoning their station-keeping and turning through 180 degrees. Within seconds they find themselves in the cod end, no longer able to outswim the net and too large to escape through the cod end meshes.

Once it had been demonstrated that trawls are as much herding devices as sieves and that fish tend to keep station on moving pieces of fishing gear, trawl designers realized that only the cod end and the parts of the net in front of it needed to have meshes small enough to retain fish. Large meshes herded just a well as small ones and generated much less drag. All these improvements were built into the net that *Tellina* was about to shoot.

Shooting a trawl is an elaborate procedure, whether done over the side, in the traditional way of the side-trawler, or over the stern, as is normal in more modern vessels. Into the sea goes the net. It streams astern until the sweeps and bridles attached to each wing of the net are fully paid out. The doors, or otter boards, are clipped on and released. The warps are paid out from the winch or winches until the net is at the required depth, or on the bottom, and fishing proper begins. In research vessels, the length of time over which the trawl is fishing tends to be shorter than on commercial fishing boats because the scientist's goal is usually to sample fish rather than to catch them in quantity. Yet, however long the tow, hauling the net, in which the actions of shooting are exactly reversed, is a time of excited anticipation.

The first indication of likely success is the number of birds following the ship. Herring and lesser black-backed gulls, kittiwakes and fulmars noisily squabble above the net as lordly gannets fold their wings and dive at small fish escaping through the meshes. At the surface, the cod end – named not after the fish but the codpiece of late medieval dress and buoyed up by the distended swim bladders of the fish in the net – furrows the wake. Some of the bolder gulls land on the cod end and peck at the trapped victims. The body of the net is pulled clear and, in modern ships, stored on a net drum. The cod end is swung into the air and its terminal knot untied. The catch cascades on to the deck and sorting begins.

The sight of a good catch tumbling about one's feet seems to awaken primeval instincts which cure seasickness quicker than anything that can be bought at the chemist's. My experiences aboard *Tellina* had made me as keen on trawl-fishing as I was on wildfowling and I was anxious to try something on a larger scale. I did not have long to wait. The results of the tomato-pip hunt had been reassuring up to a point. The seabed had stood up well to the rude assaults delivered daily by the spotless ship's company of the sludge-dumping vessel. We still had some worries about the possible over-fertilizing effects of the nitrates in the sludge on algae in the Southern Bight but, compared with what poured out of the Rhine every day, the sludge boat's contribution was a mere sideshow. What left a lingering doubt was finding dead marine animals, especially dead fish, in *Tellina*'s tiny beam trawl. Perhaps we should go back equipped to fish with gear either similar to, or larger than, that of the commercial vessels using the area to see if there really was a problem. The hint, subtly dropped in the right place, that the bed of the Outer Thames Estuary might, at times, resemble a great piscine Forest Lawn Cemetery raised the odd eyebrow in the corridors of the great. Even 'Master', Ian Graham, got to hear about it and I was not entirely surprised, shortly afterwards, to receive a letter offering me the use of the Ministry's most modern research vessel.

Corella's home port was Lowestoft, still at that time an important port for boats fishing the Southern Bight for plaice and sole, and a seaside resort whose proud boast was that it was 'First to see the sun'. That it was also the first to meet the north-easterly gales which, every so often, ripped down from the Baltic was not mentioned in the brochure. The day we joined *Corella* in the upper dock was pleasantly autumnal. The sky was overcast, but what wind there was came from the south-west, so our short passage south would be in the lee of East Anglia. We cast off with the Merchant Navy Master, Bill Craig of St Abbs, on the bridge and with the fishing mate, a tall veteran of Harry Tate's who had commanded the frighteningly cranky Admiralty trawler-minesweeper *Sir Galahad* during the war, closed up at the wheel.

Putting to sea, even in a small ship, is always a bit of an occasion. The shining white of the superstructure, the buff funnels arranged abeam like those of a nineteenth-century battleship and the hull in the full livery, inside and out, of Nelson's navy generated a huge sense of pride. As we passed through Lowestoft's swing-bridge, it was hard not to feel sorry for the harassed drivers of the queues of cars we had held up on either beam. We were off to sea in a superb new ship and they were merely off, grey-faced, to work.

As we left the harbour to negotiate the sandbanks which guard England's most eastern shore, *Corella's* bluff bows rose and fell; she was a living thing in her natural element. I left the bridge to check the security of our gear. Even on a good day, the short seas of the Southern Bight can smash jars and cover the deck with razor-sharp shards. All was well and, clear of the banks, we altered to starboard to begin our passage south. After dinner in the saloon (only warships have wardrooms) under the pale gaze of Annigoni's portrait of the Queen, Bill and I roughed out our programme. Just before dawn the following morning, we were on station in the Outer Thames Estuary. On the starboard quarter twinkled the lights of the Essex coast while, here and there, brighter flashes came from the navigational lights marked on the Admiralty Chart. We were about to undertake an echo-sounder survey to identify potential 'fasteners' or outcrops which might snag the trawling gear or worse.

The survey took all morning and most of the afternoon, recording the positions of any fasteners on the echo-sounder paper as soon as they were revealed by the trembling tip of the marker pen. The trawl-fishing grid was chosen in the light of the echo-sounder's warnings, the fishing mate's local knowledge and the objective of the cruise: to investigate the dead fish report. We then fished solidly for the next three days up- and down-tide of the sludge-dumping ground and in control areas well clear of it. Every so often we would pick a few dead fish out of the cod end, but not once when we were anywhere near the dump site. Most were whiting, *Merlangius merlangus* (L.), too small to land legally and almost certainly dumped by the small inshore trawlers we encountered scratching a living fishing for plaice, *Pleuronectes platessa* L., sole, *Solea solea* (L.), and codling, *Gadus morhua* L. Our own catches were dominated by the same species yet also included a single Twaite Shad, *Alosa fallax* (Lacepède), a rare deep-bodied relative of the herring, *Clupea harengus* L. I ate it grilled after supper, to the great disgust of the cook.

Our trawling produced one other surprise. In gutting the codling, we found their usual fish and shellfish prey but, inside the stomach of one, we came across the partially digested stomachs of five others. This enterprising cannibal had been feeding on the guts of its nearest and dearest as they were being thrown over the side by one of the small trawlers as it processed its catch. By the end of the cruise we felt we knew enough to put the mind of 'Master' at rest even if, in the process, we risked putting him off his dinner.

———

'DO YOU READ? OVER'

———

Billy Burroughs and I were in *Tellina*'s wheel-house off Spurn Head when the nearest coastal radio station, Humber Radio, called us up. *Tellina* was doing her usual best to bury her nose into the moderate head sea, but Billy had her under a light rein at seven knots and only the larger seas came aboard. In the days before mobile or satellite telephones, the only way a telephone ashore could be connected to a ship at sea was via what was called a link call. We normally had one each morning from the Marine Superintendent, at that time the genial and highly competent master mariner, Captain Aldiss. He made these calls so that we could make our SITREP (situation report) and he could pass on any new orders. A link call in the afternoon was unusual for us, and so when the bridge radio crackled out our call sign and the words, '*Tellina*, *Tellina*, *Tellina*, this is Humber Radio, do you read? Over,' we were a little surprised.

Captain Aldiss asked for me and passed on the news that my father was dying, and that he had arranged for a car to pick me up in Grimsby. Billy chose a moment between seas to put *Tellina* about and succeeded without spilling any of Jack's saucepans. It was a melancholy return to the Humber, my spirits and Billy's briefly lifted by the lovely old country march, 'Sussex by the Sea', which had been requested on the BBC radio programme we were listening to in the wheel-house. I got home in time, although my sea jersey was a little out of place among the white coats of the hospital staff.

The wide-eyed husk that remained after his valiant spirit had slipped away was not my father. Sad as I was, I immediately felt a strong sense that his vigorous younger self, the hero of my child-hood, had in some way reasserted itself. He was a man in whom the instinct of the hunter was strongly developed. Too impatient to share my maternal grandfather's love of coarse fishing, he was at his happiest walking, gun under arm, among the hedgerows and little fields of his native Buckinghamshire. How great a debt I owed him and how much he would have enjoyed the much greater sporting opportunities that were to be my undeserved inheritance.

Like many country people, my father had had little time for doctors, preferring to ignore the symptoms of diabetes mellitus until its dreadful side-effects crippled his heart and ruined his sight. But his suffering is now a distant memory, and his reborn self is ever my companion of the gunroom and the fellow 'fowler of the foreshore.

My father at the wheel of the whelk boat Elizabeth *in 1957*

My father's remains, but not his spirit, were buried along-side those of his own parents in the little churchyard at Stoke Mandeville, the parish church farthest from the sea in the whole of England.

DEATH'S DITCH

By the early 1970s, the morbid media interest in marine pollution, which had been given such a fillip by the grounding of the *Torrey Canyon*, had begun to flag. A Royal Commission on the topic had reported that the most important problems were the degradation of estuaries and, locally, of coastal waters and the accumulation of so-called conservative chemicals such as chlorinated hydrocarbons and polychlorinated biphenyls (PCBs) up the food chain. Conservative chemicals accumulate in this way because they are not readily broken down in the tissues of plants and animals. Most are soluble in fats, however, and so when one organism eats another, some of the contaminant chemicals in the fatty tissues of the prey are transferred to those of the predator. As more and more contaminated prey are eaten, so the concentration of the chemicals builds up in the body of the predator. The mere presence of contaminants is not, of itself, evidence of biological harm but, given Man's place at the very top of the food chain and the known effects of some chlorinated substances on such factors as eggshell thickness in birds, the Royal Commission was right to be cautious.

My own interest had been in the tangible effects of pollutants in coastal waters and I was beginning to run out of problems.

I reckoned without the activities of one Dr Perkins, a university teacher whose well-publicized observations on the incidence of skin disease in the fish of the north Irish Sea had convinced him that they were, in some way, the victims of pollution. Once again, 'Master' and his officials were concerned. There was a distinct danger that the Minister might be asked a question which he was able neither to answer nor to avoid by the traditional 'Yes, Minister' ruse of answering a question he had not been asked.

The genteel Welsh biologist Ivor Rees and I had been to Dr Perkins's back yard not long before in the Lancashire and Western Sea Fisheries Committee's Patrol Vessel *John Beadsworth* and, later, in the shiny new oceanographic research ship *Prince Madog*, operated by the Menai Bridge Laboratory of the University of Wales. This little armed reconnaissance had been in connection with local worries about the fate of Liverpool's sewage sludge. So far as Ivor and I could see from looking at the seabed, the sludge was well dispersed by the strong tides and often turbulent seas of the north Irish Sea.

We did have a worry, though. If you put organic wastes into the sea, you are fertilizing it with nitrates and phosphates, nutrients which can, under certain conditions, encourage the growth of marine algae. Whether or not it is possible to estimate the effects of these wastes on the production of algae depends, among other things, on the relative scale of the dumped input: much larger quantities of nutrients are usually present, but these tend to have been naturally recycled from the seabed or run off from the land. The degree of sunlight penetration also matters a great deal. Often in the shallow and turbulent waters of the north Irish Sea, the water is so turbid that high levels of nutrients can be present without having much of a fertilizing effect on the algae. Sometimes, though, during periods of settled, warm weather, the sea clears, the sunlight beams into the depths and the algae do so well that blooms of less desirable species develop.

The less desirable alga in the north Irish Sea was a colonial form called *Phaeocystis*. The colonies take the form of blobs, in the surface of which the individual algal cells are embedded. In life they are rather beautiful things, microscopic jewels voyaging together in a soft ball of jelly but, when their time is up, the jelly breaks down. Waves breaking on shore convert the resulting mess into great lumps of foam. The press, still in the residual grip of pollution hysteria, chose to connect one of these foam fests with Dr Perkins's observations on fish disease and the almost certainly natural deaths of large numbers of sea birds following a storm. They christened the north Irish Sea 'Death's Ditch'. It was time to check over my sea kit.

By this time I knew Dr H. A. Cole of Lowestoft well enough to cut a few corners and ask directly for the services of FRV *Corella*, newly refitted after a fire which had started under mysterious circumstances in the new fishing mate's cabin. Permission was granted at once and I shortly found myself on the bridge with Bill Craig as we once again held up the traffic at Lowestoft. The shipping forecast was grim: 'There are warnings of gales in Dover, Wight, Portland, Plymouth, Sole, Lundy and Fastnet.' The heavy weather was to be from the south-west, just what we did not want during our long steam down-Channel and round the south-west tip of England. It was none too comfortable in the Southern Bight of the North Sea either and some of my scientific colleagues began to suffer severely. By the time we got to Rye Bay off the coast of Kent, we were in a lee which gave sufficient shelter for us to shoot the trawl and collect a large control sample of flat fish for later comparison with the allegedly grisly victims of 'Death's Ditch'.

It was to be but a short respite. We were off the coast of Sussex when the gales hit us in earnest. The seas had been building up steadily in response to a storm far to the west – the dreaded 'dog before its master'. Now we eased down to around nine knots, *Corella* pitching horribly, the great stern gantry swinging the hull up and down like a giant pendulum. What a way to travel! Even the pre-lunch pink gin, at that time nine shillings (45p) a bottle from the bonded store, lost its attractions. We eventually clawed our way to Falmouth, the wind by now 'Severe gale nine gusting to storm ten'. We hove-to overnight, the bridge watch holding *Corella*'s bluff head up to the sea and Bill Craig appearing at intervals to confirm our position relative to the lee shore (land downwind of the ship) represented by the rocky coast of Cornwall.

'You're a stormy petrel, Dr Shelton,' observed the poor cook the following morning. He had a small cut over one eyebrow and had spent the night climbing back into his bunk in the forepeak, the most uncomfortable place in the ship. He and a number of the other hands would like to have spent the night alongside in Falmouth, but Bill was not keen on the crew having a convivial 'run ashore', knowing as he did that the following morning we would be entering the maelstrom once more.

We had not yet seen the worst. By the time we were off Land's End, the wind was gusting to force eleven. We had also altered course so that, instead of heading directly into the great seas, we now had them just abaft the port beam. *Corella* had been fitted with a stabilizing system in which two large tanks arranged athwartships were connected by a narrow neck. As the ship rolled, so the tanks exchanged their contents via the narrow neck. The effect was to damp the roll. The price paid was sometimes an unnatural delay in the response of the ship to the sea and this was one of those occasions. It was impossible to stand up without a secure handhold and I was thankful that my battered scientific colleagues, one of whom had never been to sea before and was finding it all rather trying, had turned in. Slowly, imperceptibly, as we gained the lee of Ireland, *Corella's* motion became easier. At the same time the wind velocity started to fall as the gale blew itself out.

At last the cook was able to provide more than cold food and soup and we could sit down together in the saloon. Struggling against the gales had burned up a lot of diesel oil and our first priority, once we were in the north Irish Sea, was to go in to Douglas in the Isle of Man to refuel. At last the crew had their run ashore, an opportunity to refresh themselves of which some took full advantage. A few were still not quite with it when we put to sea on the following morning. Among them was the new fishing mate, a plump son of Suffolk still in his late thirties.

We started the first of our grid of trawling stations within three miles of the Isle of Man coast, a place denied at that time to commercial vessels. Round the corner came a minesweeper of the roll-on-wet-grass 'Ton' class, a member of the Fishery Protection Squadron. It was the proud, if unstable, charge of a young lieutenant commander enjoying his first sea command. 'Stop your vessel at once,' flashed his Aldiss lamp, after he had failed to raise us on the radio. 'What do you think we should do?' I asked the young fishing mate. 'He obviously does not realize that we are a research vessel.' 'Ignore the punt' (at least that's what it sounded like) was his succinct, if unrealistic, advice – advice I was unwilling to take, given the proximity of the minesweeper's Bofors gun. I called Bill, an expert with the Aldiss lamp, and we sent the would-be flag officer on his way. Later on, we started to pick up pieces of a Buccaneer carrier-borne strike aircraft in the trawl. One had exploded in midair during a naval exercise some weeks before and it is possible that our grey-funnelled chum had been aware of the fact.

Not all of the crew credited the new fishing mate with the qualities he felt his elevated status demanded. Given that the grid of trawling positions I had given him was selected for scientific sampling reasons, rather than to catch fish in quantity, it was inevitable that often the catches would be light. I had also asked that tickler chains be rigged in front of the trawl's foot-rope to improve our chances with the flatfish which were our main quarry. Tickler chains are rigged in this way to disturb fish and shellfish buried in the sediments ahead of the net, and one of their undesirable effects was to dig up a lot of dead queen scallop shells and other debris, all of which was duly emptied out on to the deck. Among the Suffolk crew we had one experienced trawlerman from Hull who was consequently nicknamed 'Yorkie'. He and I were sorting the results of the last haul into fish boxes when, in exasperation at all the shell, he looked up at the fishing mate in the wheel-house and quietly observed that 'Any fooker can catch shite'.

Certainly, the relationship between the fishing mate and the deck-hands could have been better. One of the most senior, a Harry Tate's veteran who had attained the temporary rank of lieutenant commander during the war, was nicknamed 'Squeaker'. He was, by this time, a frail figure who loved the sea but was sometimes ill ashore, occasionally quite seriously. He came on to the bridge to take over the wheel and casually observed to the fishing mate that he had 'nearly died' on his last leave. 'Trouble wi' you, Squeaker, you never make a fuckin' job of it' was the laconic East Anglian reply. But this was to be the fishing mate's penultimate cruise aboard *Corella*. On the next deployment, he assaulted my successor while he was sitting at his desk and thereby earned immediate dismissal.

Back in the stable world of the laboratory, we could forget about the shipping forecast and analyse our results. In the fish we examined, we found three main sorts of lesion: cysts caused by the common virus disease *lymphocystis*, bacterial ulcers and fin damage. All three sorts of lesion were more common in the north Irish Sea samples and, in both those and the Rye Bay fish, *lymphocystis* and fin damage in flounders, *Platichthys flesus* (L.), were the main problems. There is not much of a market for flounders, perhaps because they often carry an earthy taint reminiscent of the geosmin sometimes found in freshwater fish. Fishermen call them Morecambe Bay halibut and usually discard them, their skins no doubt abraded by contact with the net. The trauma thus produced, and the tendency of flounders to spend part of their lives in brackish and even fresh water, may well predispose them towards skin diseases. Whether the differences between the Rye Bay and Irish Sea results reflected the higher fishing intensity in the Irish Sea or some other factor, we were not able to establish. What we were quite sure of was that the journalists' appellation, 'Death's Ditch', was complete nonsense.

Nonsense or not, the news media had long since moved on to some new horror, real or imagined, and devoted no column space to the results of our sober assessment. It is, I suppose, inevitable that journalists cherry-pick scientific results for those few reports that sell newspapers: nothing seems to appeal more than a good scare story or, better still, dark rumours that one has been covered up. Irritating and superficial as such press coverage can be, over the years I have learned to value it as one of the very few ways in which science can compete for public attention with the antics of footballers and pop stars.

———

SHRIMP SOUP AND CRAB BAIT

———

Much as I had grown to like working at Burnham, and engaging the widgeon on the Blackwater foreshore, I still missed Scotland. An interesting post came up at a Scottish university and, following a clandestine application, I was invited for interview. My second son, Neil, was born during the twenty-four hours I was away, entering the world in a fraction of the time taken by his elder brother, who had been in no hurry to meet either of his parents for the first time. Fortunately, the Scottish university wisely decided to look elsewhere – I say fortunately because not long afterwards I was invited by Dr H. A. Cole to assist him in his new appointment as Controller of Fisheries Research and Development for the United Kingdom. It was an appointment, based in Lowestoft, which was to mark the last serious – but regrettably unsuccessful – attempt to fulfil Sir Alister Hardy's dream and weld the country's fishery research facilities into a single, efficient organization.

I was beginning to tire of dousing the brush fires of the marine pollution industry, and besides, the Lowestoft post had two other great attractions. The first was that Cole was due to retire before his appointment as Controller ran its three-year course: his successor was likely to be based in Scotland and my hope was that I would duly be transferred north to work with that successor. The second was the opportunity to spend some time in a laboratory with a worldwide reputation where many of the key principles of fish-stock assessment and management had first been worked out. I had not come, however, from Burnham to Lowestoft empty-handed. I still had the work on the Irish Sea flatfish to write up and publish, and I had a couple of new projects to pursue.

One of my colleagues at Burnham, the world-class yachtsman Peter Warren, was an expert on the shrimp fishery of the Wash, where hardy Lincolnshire longshoremen pursue the brown shrimp, *Crangon crangon* (L.), with beam trawls. Brown shrimps are stoutly armoured little creatures which spend a lot of their time partly buried in muddy sand. They are highly specialized crustaceans living a very different sort of life from the lobsters and crabs to which they are distantly related. Their great survival trick is their ability to shelter from predators by rapidly burying themselves so that only their eyes and two pairs of antennae are exposed. But fine muddy sand is still abrasive stuff to be dashing in and out of, and brown shrimps cope with this by having a much tougher shell than their pink shrimp relatives. Sand also takes a lot of shifting, even though the individual grains weigh less underwater than they do in air. Once again, the brown shrimps have a trick up their sleeves. The bundles of myofibrils that make up the muscles of crustaceans are closer-packed in *C. crangon* and are correspondingly stronger as a result. It is the high density of these muscles that gives brown shrimps the meaty texture and intense flavour so prized by con-noisseurs of shellfish.

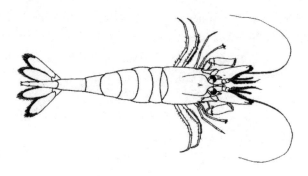

The brown shrimp, Crangon crangon *(L.)*

When Peter Warren showed me some of his live specimens, my earlier interest in crustacean sense organs resurfaced. I no longer had the means at hand to record electrical impulses from nerve cells, but there was nothing to stop me from watching the animals in action and looking in detail at their structure. A few drops of mussel juice from a pipette suggested that the chemical senses played an important part in their food-seeking behaviour. Like lobsters, shrimps have specialized 'aesthetasc' setae on their antennules (the first pair of antennae) which are exquisitely sensitive to chemical stimuli. There is a double row of these specialized organs on each antennule. No one really knew how they worked, only that there appeared to be a small pore at the tip of each one through which chemical stimuli were assumed to pass.

One evening, after most of the staff had gone home and the telephones had ceased to ring, I thought it would be rather fun to look at the *Crangon* aesthetascs under the microscope to see the terminal pore myself. What I did see astonished me. In many of the specimens, strong filamentous growths projected from the aesthetasc tips, growths which appeared at first sight to have the cellular structure of blue-green algae. The shrimps still seemed perfectly able to scent their prey, but how could this be if the terminal pores of their aethetascs were stuffed up with algal filaments? Surely it would be like trying to nose a vintage claret with a large cabbage growing out of each nostril.

My first paper on the subject merely reported what I thought I had seen under the light microscope. The response, from a Dr Brock, a microbiologist who knew far more about the subject than I did, could not have been more helpful. He immediately identified the blue-green alga as the filamentous bacterium, *Leucothrix mucor* Oersted. Unlike a true alga, this organism is unable to use sunlight to help it build up its body from simple components, but depends upon a ready source of complex organic molecules, just as we and all other animals do. Where better to find such juicy molecular titbits than at the front end of a feeding shrimp? So much for why the filaments were growing where they were – but just how could the shrimps still smell normally despite their blocked pores?

I had taken the observations as far as I could on my own and needed the help of collaborators with more powerful microscopes to look at the pores. I also needed a regular source of shrimps. My brother Peter, by this time teaching at the University of Leicester, and my old St Andrews friend Sandy Edwards provided the microscopical muscle-power, while ten consecutive monthly samples of shrimps were supplied by Skipper F. J. Symonds of Great Yarmouth, the father of one of my former colleagues back at Burnham. Under the scanning electron microscope, we could see that the aesthetascs were flexible segmented structures, the last segment of which narrowed to a nipple-like point. There was no sign of a pore in the tips of the intact setae, a conclusion we later found out had also been arrived at by a Dr Snow, who had been looking at the antennules of hermit crabs.

Like Dr Snow, we found that a high proportion of the aes-thetascs had lost their tips. This was especially true of shrimps caught in the winter, when they can spend a long time between moults and there is a greater opportunity for damage to delicate structures and for infection to become established before the old shell, along with all the organisms growing on it, is shed. Indeed, the appearance of a terminal pore seemed to be the result of damage, the site of which provided a ready foothold for infection by *Leucothrix mucor* Oersted. However, unlike the other setae borne by brown shrimps, the cuticle that forms the shaft of an aesthetasc is clear and unusually thin. This is because it does not contain the brown tanned protein that stiffens and thickens the other setae. As a result, stimulatory chemicals are able to pass freely through the side of the aesthetasc, which can therefore go on functioning as a sense organ whatever may be growing out of its tip. Thus the shrimps were still able to sniff out their prey.

It was to be some years before my world and that of shrimps would overlap again and then it would be the pink shrimps, *Pandalus borealis* Kröyer, living in the Farn Deeps off north-east England whose lives I would glimpse through the lens of a CCTV camera. More lightly constructed and lobster-like than the brown shrimps of the Wash, they sought shelter and protection, not by burying themselves but by living in the lee of the large sea anemones known to fishermen as 'horses' arseholes'. The pink shrimps also make use of another survival trick, one which exploits the fact that it takes a great deal more energy to produce eggs than spermatozoa. It is therefore often possible for male animals to accumulate enough surplus energy – usually as fat – for them to achieve sexual maturity earlier in their lives than females. Pink shrimps maximize their chances of leaving offspring by maturing first as males and the following and any subsequent years (they rarely live for more than three) as females. In brown shrimps, the sexes remain separate throughout, perhaps because building up the muscle mass they require to shift sand leaves no spare energy to fuel two different bites at the reproductive cherry. What remarkable lives shrimps lead, and how little thought we give the matter as we eat them.

Little detective stories, like the tale of the brown shrimp's nostril hairs, keep scientists sane and amused, but they do not put more money in the fisherman's pocket. One related area of work which had the potential to do just this was the development of an effective artificial bait for the lobster and crab fisheries. At certain times of the year, bait shortages can cause real problems, with fishermen reduced, at times, to using such expensive alternatives to salt fish (for lobsters) or fresh fish (for crabs) as roughly perforated tins of cat food.

Before leaving Burnham, I had been introduced to the marine biochemist Dr Sandy Mackie, who was based in the Fisheries Biochemical Research Unit in Aberdeen. I explained the bait problem to Sandy and he suggested that we try and analyse the main soluble compounds found in squid flesh, one of the most effective baits for both lobsters and crabs. He started by separating the fat-soluble compounds from the water-soluble ones and sent me samples of each extract. I had no easy access to lobsters so instead used a miniature version, the delicate estuarine shrimp, *Palaemonetes varians* (Leach), found in every pool on the salt marsh.

The little shrimps were strongly attracted to an extract of whole squid and to the water-soluble extract from which all the fatty substances had been removed. They showed little if any interest in the fat-soluble material. We could, therefore, be reasonably sure that, so far as crustaceans were concerned, the widespread belief that the attractiveness of baits resides in their oil content was incorrect. A further surprise came when Sandy sent me different fractions of the water-soluble extract of squid.

We had expected that, by splitting the extract up by the mysterious processes known only to biochemists and 'other honest men', it would be possible to concentrate the attractive chemicals and identify them. But the shrimps had not read the script. No one fraction was as attractive to them as the whole water-soluble extract, the attractiveness of which we could match only by re-combining the fractions. The important compounds from the shrimps' point of view were amines, amino-acids and nucleotides, substances which are present in complex mixtures in the haemolymph (the equivalent of our blood) in many invertebrate animals, including the prey of shrimps.

With the help of a gifted undergraduate visitor, I was able to confirm the results we had obtained with the shrimps in a separate set of experiments with shore crabs. We also tried mixtures of the pure compounds in the proportions in which Sandy had found them in the squid. Once again, the complete mixture always outperformed any one fraction. On this basis, we concluded that the manufacture of an artificial bait for the lobster and crab fisheries would be prohibitively expensive and, thanks to the hard-wired powers of discernment built into the nervous systems of crustaceans, might well prove less attractive than the real thing.

Food is the raw material for survival, growth and reproduction, so getting enough of it is important for all animals. It is not just a matter of quantity, though. Matching food to a species' particular dietary needs demands selection by the senses, especially in the case of benthic or seabed-dwelling crustaceans, the chemical ones. It is therefore not surprising that an olfactory selection system honed by over a hundred million years of evolution cannot be fooled by simple cocktails of chemicals. Such a conclusion would also be no surprise to Egon Ronay.

PREDATORS AND PREY

The world reputation of the Lowestoft laboratory rests largely on the contribution it has made to our understanding of the effects of fishing on the stocks of fish in the sea. Although they are rarely expressed in similar terms, the problems posed for fishery science are no different, in principle, from those faced by scientists studying other relationships between predator and prey. For both, as for all living things, the primary imperative is reproduction, and for this to be possible, the individual predator and prey have each, through their food, to acquire energy resources over and above those they need to maintain their bodies and to grow.

When food is plentiful, it does not take long for animals to build up enough surplus energy, stored mainly as fat, to channel some of it into reproduction. The more plentiful the food and the easier it is to obtain, the sooner sexual maturation can be triggered by environmental cues like the rapid change of day-length around the spring equinox. Our own species provides an excellent example, for plump girls tend to have an earlier puberty than their slimmer sisters. Atlantic salmon respond to better feeding opportunities in the same way. Fish of comparable parentage mature earlier when fed *ad lib* in a fish farm than when they have had to find and catch their own food in the wild. The other side of the coin is that, when food is hard to come by, not only is there no reproductive surplus but growth-rate also suffers. Reproduction is delayed and the fish remain small for longer. Small fish are vulnerable to a wider range of predators than large ones. They also swim more slowly and so are easier to catch. When eventually the survivors are ready to breed, there are fewer of them to found the next generation.

Both predator and prey are constrained in their respective growth, survival and reproductive rates by the same iron discipline imposed by feeding opportunity. For instance, puffins depend for their early survival on the availability of young sand eels. In years when the numbers of young sand eels are low, the next generation of puffins suffers accordingly. To this extent, predator populations are ultimately controlled by the abundance of their prey. Feedback loops of this kind are the principal reasons why predators which have evolved alongside their prey tend not to over-exploit them.

In reality, of course, the connection between predator and prey is not between two species, but many. The complex relationships that bind the lives of predators and prey together make up what ecologists call a food web. To take one example, salmon feeding in the North Atlantic are not dependent upon any one species of crustacean or young fish. Like many species, they are opportunistic predators, feeding on whatever is most available at the time. Feeding in this way confers a degree of stability on the salmon population because reductions in the availability of one food organism may be balanced by increases in that of another. Krill may be abundant when amphipods are scarce and lower numbers of post-larval sand eels may be balanced by increases in the numbers of juvenile blue whiting and five-bearded rockling.

In the same way, the marine birds, larger fishes and marine mammals that prey opportunistically upon salmon are not exclusively dependent upon it for their own growth and survival. Thus the marine food web is made up of countless predator–prey relationships which vary in strength as some species wax and others wane. When Man, the terrestrial predator, acquired the means to become an aquatic one, he also took his place near the top of the marine food web. At first, and for thousands of years thereafter, Man, the subsistence fisherman, dealt in the same currency – energy – as his fellow predators, and his needs were as readily accommodated. But the rules changed dramatically with the invention of the fishing vessel, and especially the fishing vessel powered by a steam or diesel engine. The new currency was money, and the fishing vessel itself became the predator. The finely balanced feedback loops linking predators to prey were broken for good. Fishery science would one day emerge as their clumsy and inadequate substitute.

The effects of these revolutionary changes were directly comparable with those following on from the introduction of an alien predator like mink to England or hedgehogs to North Uist. The effects were not as immediately drastic as those felt by water vole preyed on by mink, or ground-nesting bird populations by hedgehogs, but felt they were, and first by the men on the fishing vessels. After a period of unprecedented success, they began to notice that they had to work their boats harder to maintain their catches, especially of the larger fish which fetched the best prices. In the yet to be developed language of fishery science, their catch per unit effort (CPUE) had fallen and there had been an attenuation in the age-structure of the stock. Scientific opinion, as represented in the late nineteenth century by Thomas Huxley and William McIntosh, was dismissive, but Walter Garstang, director of the Lowestoft laboratory in the early years of the twentieth century, took a different view. It was clear to him that the new fleets of powered fishing vessels could have a material effect on the stocks of fish, and that there was a real danger of over-fishing them.

What, though, does over-fishing really mean? To McIntosh, it meant threatening the future supply of young fish by catching too many of their parents. He was convinced until the end of his life that it would take a lot more than a few steam trawlers to achieve such a result. Although some fishermen were not above deploying McIntosh's argument against rivals when it suited their case, their immediate concerns were primarily economic ones. Having to steam longer and farther from port to maintain catch-rates used more coal or oil, and small fish made less per pound than the larger ones they could remember from the not so distant past. Deeper thinkers began to realize that the economic effects being felt by the fishermen did not necessarily threaten recruitment – the term scientists apply to the next generation of young fish – but, if pressed too far, were evidence of wastefully high levels of exploitation. Fish, which in the past had had ample time to grow large and valuable, were being caught before their growth potential was fully realized. Were the fishermen not right in regarding this as a form of over-fishing?

Dr E. S. Russell, the ambitious Scot who directed the Lowestoft laboratory for most of the interwar years, pointed out that, if a stock of fish were treated as a single entity, its total weight at any one time depended upon the relative strengths of the forces tending to augment or decrease it. He expressed this idea in the form of a simple algebraic expression. The plus side was represented by the weight of young fish recruiting to the stock (R) each year and the annual growth in weight of the fish already in it (G). Annual deaths of fish due to fishing (F) and natural factors (M) represented the negative forces.

As the fishermen's experience had shown, one of the effects of increasing the negative forces – say, by increasing the intensity of fishing for a few years – is to thin out the accumulated stock of slow-growing older fish and increase the proportion of faster-growing younger ones. To this limited extent, Russell postulated that fishing at moderate levels helped to generate the production upon which it depends – a process his successor Michael Graham was to compare with the effects of mowing a lawn. So far as the recruitment of young fish was concerned, empirical evidence suggested that moderate levels of fishing did not threaten it. Later, it was realized that the reductions in the numbers of newly hatched larval fish, caused by the reduced numbers and sizes of their parents, tended to be balanced by the greater survival which the young fish enjoyed prior to recruitment when, being less numerous, they had access to better individual feeding opportunities. This is the forgiving mechanism by which stocks of fish compensate for the additional losses created by fishing. Experience with many species was to show that the numbers of adult fish could vary between wide limits before the number of eggs they laid and fertilized directly limited the numbers of young fish appearing in the fishermen's nets.

Thinking along the simple lines propounded by Russell was readily understood by both politicians and fishermen. It was to give rise to what became known as the surplus production approach to modelling fish populations. The underlying assumption here is that, in the absence of fishing, the total mass of the stock will rise to some limit set by the capacity of the marine environment to support it. As the accumulated stock is fished down and compensatory increases in the individual survival rates of the youngest fish, and in the growth-rates of the others, kick in, so additional surplus production is created. Thus the initial fall in catch-rates does not, in itself, constitute over-fishing. Over-fishing begins when too few fish survive to realize their potential for growth. Such 'growth over-fishing' is both economically and biologically wasteful, yet it does not threaten the strength of succeeding generations. However, if so many adult fish are caught that the recruitment of young fish is restricted by the size of the parent population rather than the capacity of the environment to support those young, a state of 'recruitment over-fishing' is said to exist. Alone, or in combination with a deterioration in the environment of the young fish, recruitment over-fishing can cause the complete collapse of stocks.

After the Second World War, great changes took place in the structure of the fishing industry. These included the opening-up of the Barents Sea to British vessels and the development of dangerously more effective methods of detecting and catching pelagic fishes like herring. Faced as the Lowestoft scientists were with increasing demands for advice on how fish stocks could best be managed, the limitations of the superficially attractive surplus production approach became increasingly apparent to them. Stocks of fish are not corporate entities but assemblies of different individuals.

As a consequence, models which neglected this truth often proved to be misleadingly unrealistic because they took no account of detailed changes in the structure of the fished populations or in the underlying productivity of the environment on which they depended. With the enthusiastic support of its new director Michael Graham, two young scientists at Lowestoft, Ray Beverton and Sidney Holt, pioneered an alternative approach to modelling the effects of fishing in which the dynamics of the stock of fish are considered in relation to the fate of individuals. This analytical approach to modelling, as it came to be called, was demanding of data and, as a result, expensive to put into practice. However, as a way of describing the lives and ultimate fate of successive generations of fish, it took some beating.

The older fish become, the larger they grow but, as the years pass, the number of fish in each generation is steadily reduced by fishing and natural predation. As a result of these competing influences, the total weight of a generation of fish rises to a peak and then tails away to nothing as the last fish dies. The new analytical models were able to describe how increasing the death-rate through fishing influenced the point at which the peak was reached. They showed that, to make the most of each generation of fish, it was important neither to fish down the numbers of fish too soon nor to miss the boat by waiting so long that most of the fish had died of natural causes.

The size at which fish first become vulnerable to capture is ultimately limited by the mesh size in the catching part of the net but, for fish large enough to be retained, the chances of being caught rather than dying off naturally depend upon the rate of fishing. The beauty of the analytical approach was that it was able to indicate which combination of fishing effort and size at first capture should give the best yield; the general rule being that the greater the fishing effort and the lower the rate of natural loss, the larger the size at first capture needs to be. What this means in practical terms is that, if fishery administrators wish to alter the numbers of fishing vessels or the mesh sizes of their nets, the effects of their actions in getting more or less out of a fish stock can be predicted with fair confidence.

Although calculations of this kind provide valuable theoretical guidance on how to obtain the best yield from a fishery, they do not, of themselves, predict what the absolute yields will be. That is because they do not incorporate estimates of the numbers of young fish that recruit to the fishery each year. The output of such partial calculations is therefore described by fishery scientists as a yield-per-recruit assessment and not an assessment of the total yield that might be expected. For many species, the numbers of young fish which enter the fishery each year vary within very wide limits. The range of variation is especially wide among the members of the cod family where, in the case of haddock, successive brood years may sometimes vary by a factor of 100. Such enormous differences are believed to reflect the degree to which there is a match – or mismatch – between the time at which the eggs hatch and the feeding and survival opportunities of the young fish in their first weeks of independent life.

Annual variation in the numbers of young tends to be less when the juvenile phase includes a period when the availability of a territorially limited habitat constrains the numbers of young that can be supported. Thus production of young plaice in the North Sea appears to be partly limited by the availability of shallow sand flats in areas like the Friesian coast. For lobsters, the limit is provided by reef habitat and, for Atlantic salmon, by the size and quality of their rivers of origin. Despite all the empirical evidence of the difficulty of the task, valiant efforts have been made to describe recruitment processes mathematically from the starting assumption that, in the longest term, recruitment is some function of the size of the parental stock. However, the fact remains that, for many species, the only short-term way to predict recruitment is experimental fishing by research vessels using nets with mesh sizes small enough to retain fish just before they reach the stage at which they appear as recruits in the commercial fishery.

Nevertheless, all of these attempts by the Lowestoft scientists to use simple mathematical models to help describe the effects of fishing, to assess the need for management measures and to predict their effects, have done much to provide a quantitative basis for managing the fisheries. In the light of all this practical scholarship, and of developments to try and address the difficult problem of managing several species together, it was already clear, over twenty years ago, that the more valuable species such as cod, haddock, plaice, hake, herring and mackerel were already fully or over-exploited.

It is commonly believed that this evidence of failure is the fault of fishery administrators and politicians unwilling to take heed of scientists' warnings. There is much truth in this assertion, but it is not the whole truth. Helpful as the scientists' models were, too often their credibility was harmed by obvious deficiencies in data, these sometimes exacerbated by disagreements between scientists based north and south of the Scottish border. It could also be argued that at times scientists and administrators might have made greater use of the relationship between the scarcity of the more desirable species and their unit value. The scarcer fish like cod and hake become, the more they attract a premium at the market. In the same way, large fish attract more money than small ones and, when quotas are limited, it can make short-term economic sense to discard very large quantities of small fish in order to select a smaller number of large specimens of the same or other species. Fishing for money rather than for fish *per se*, is one of the main factors that distinguishes a fishing vessel and its crew from a conventional predator and enables it to go on depleting a resource at levels which make no biological sense.

Faced with the depletion of many of the home-water stocks, and with only limited opportunities to take the robust corrective action required, it was natural that the Fishery Departments should have wished to look beyond the traditional fishing grounds. International initiatives to extend fishing limits to 200 miles from the coast would soon deny most distant-water grounds to British vessels. Only the deep-water resources on the edge of the continental shelf to the west of the British Isles remained largely untapped, and I was shortly to have the privilege of seeing them at first hand.

––––

THE MONSTER HUNT

––––

Fish are like lawyers; they fatten on the misfortunes of others. Most are carnivores but, wherever they sit in the food chain, they are ultimately dependent on complex organic compounds generated in the cells of plants. Although there are a few marine flowering plants like the eel grass, *Zostera marina* L., most marine plants are algae. Of these, by far the greatest bulk live in the surface waters of the sea as phytoplankton – literally, plant drifters. They have three requirements: a source of nutrients, sunlight and a temperature regime that suits their internal biochemistry. The temperate seas over the continental shelf, on which the British Isles sit, are able to support high levels of algal production. Winter storms and temperature fluctuations bring nitrates, phosphates and other nutrients present at depth up into the surface layers. Here, during the following growing season, beginning in the spring, algae thrive and provide food for the swarming hordes of zooplankton which directly, or via links higher up the food chain, sustain the fishes of commerce.

The production of phyto- and zooplankton in the surface waters of the open ocean obeys the same rules but, except in areas where deep water, rich in nutrients, wells up from below, is limited by the lack of such fertilizer. Levels of production at the edge of the shelf seas, the so-called continental slope, tend to be between those of the continental shelf proper and the open ocean. Some of the fish that live on the slope depend indirectly on dead material raining down from surface waters. Others prey upon so-called bathypelagic species, like the lantern and hatchet, fishes which feed in surface waters at night but return to the depths during the day.

Lantern fishes are common throughout the world's oceans. Many of them look rather like the small Arctic charr found in the deep lakes of the Northern Hemisphere. Like the Arctic charr and other relatives of salmon and trout, lantern fishes have a small fin in front of the tail known as an adipose dorsal fin. Adipose, as anyone familiar with the modern fad of dieting will know, just means fatty. In the case of the dorsal fin, it signifies that it is made up of fat and connective tissue and contains none of the finger-like rays that stiffen the fish's other fins. This difference in the structure of the adipose dorsal fin is interpreted by some students of the evolution of fishes as evidence of its having been evolved separately to fulfil an important function. Exactly what that function is has never been tested experimentally, but my former colleague Dr Clem Wardle, a world authority on the swimming performance of fishes, believes adipose fins are there to reduce drag when fish are swimming hard. This they do by acting as foci from which vortices spin off, just as they do from the wingtips of aircraft.

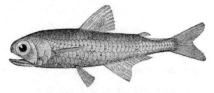

A Victorian drawing of a lantern fish

What gives the lantern fishes their name is not, of course, their fins but their light organs. Unlike politicians, the light organs of lantern fishes have the near miraculous ability to generate light without heat. They are mainly distributed along the underside of the fish and one of their functions is to render the fish less visible from below against sun- and moonlight shining down from the surface. Most lantern fishes are no longer than five or six inches and are silvery with a dark cast. They are often found in the same hauls as the distantly related hatchet fishes. The mirror-like bodies of hatchet fishes are much foreshortened and they look like so many bright silver crowns when tipped out on to the deck. They also use light organs to conceal themselves from below and share the lantern fishes' adipose fin, albeit in a much elongated form. On dark nights, hatchet fishes can come surprisingly close to the surface – so much so that on one occasion my friend Peter Foxton surprised a shoal of them off the coast of North America, feeding on ladybirds which had been blown offshore.

For all their enormous abundance throughout the world, lantern and hatchet fishes have little direct significance for Man, save for the problems which the sound reflected from their swim bladders creates for submariners and the crews of the ships that hunt them. We were after the bigger game that lived not in mid-water but on and just above the bed of the continental slope. Getting one's hands on such fish in quantities sufficient to judge whether they might form the basis for a viable fishery is no easy task. The shelf edge to the west of Scotland and Ireland is a long way from the nearest harbour, so the first requirement is a ship which can live in the worst weather the North Atlantic can throw at it. The second is specialized expertise in the handling of trawl nets at depths as great as 700 fathoms or 4,200 feet. So far as the ship was concerned, the Lowestoft laboratory's steam side-trawler *Ernest Holt*, veteran of Iceland, the White Sea and Bear Island, had just come to the end of her life as a research vessel. She was replaced by Geoffrey Trout's masterpiece, the research stern trawler *Cirolana*, which was *Corella* writ large and without compromise.

Painted completely white like the Royal Navy's hydrographic survey vessels *Hecla* and *Hydra*, to fit her, if required, for service in the tropics, *Cirolana* incorporated the very best of commercial practice with the modifications required for a wide variety of fishery and oceanographic tasks. In sharing *Hecla* and *Hydra*'s livery, *Cirolana* was following in the most distinguished of footsteps. An earlier *Hecla* had taken part in the 1819 expedition to seek the

North-West Passage, and in 1868 a former *Hydra* was instrumental in developing better methods for sounding the deep ocean.

Cirolana's 235-foot steel hull, with fine entry but good forrard buoyancy, ensured that she had excellent sea-keeping qualities. She was equipped with an improved version of the passive stabilization system fitted to *Corella*, and another lesson from that ship's experience, a bow thruster, guaranteed precise station-keeping. The main engines were mounted on a massive sprung raft to ensure that, so far as possible with this method of propulsion, vibration was not transmitted to the hull and thereby out into the sea or into the ship's laboratories. Finally, a vital necessity in fishing vessels, there was an excellent view of the working deck from the bridge. The overall result was outstanding and, like the Concorde supersonic airliner and so many other good pieces of engineering design, it looked as good in the early 1970s as the *Ernest Holt* had in the late 1940s.

If we were to be well served by our ship, we were equally well-equipped with specialized fishing skills. My fellow wildfowler John Bridger, a senior net specialist from the laboratory, and George 'Schnozzle' Argumont, fishing skipper of *Cirolana*, had combined their experience to design a trawl that would cope with most of the hazards it would be likely to encounter operating at depths of up to 700 fathoms. Heavy doors and strong metal floats (even some of the strongest are inclined to implode at the great pressures prevailing on the slope) on the head line served to keep the trawl open; great round bobbins, eighteen inches in diameter, were fitted to the foot-rope to help roll the net over obstructions such as rocky outcrops and colonies of cold-water corals. It was a tribute to John Bridger's diplomatic skills that the details of the net and how it was to be worked had been agreed in advance with Schnozzle. The latter had been a Don (elite) skipper in his days as a commercial fisherman on the distant-water grounds, but his directness of manner had not always been appreciated by his crews and he was once seriously assaulted ashore. Eventually, he had had difficulty in persuading deck-hands that his great skill at finding and catching fish fully made up for the unsubtlety of his personal qualities. As he put it, 'Ah never thought t' day would coom when ah'd be workin' for fookin' Ministry.' Well, 't'day' did come and, under the firm control of *Cirolana's* Merchant Navy Master, Captain Tom Finn, a survivor of a Japanese prisoner-of-war camp and a giant among men, the Ministry was greatly to benefit.

Our exploratory fishing on the slope was to take place as the second leg of a research cruise which had begun to the north of the Shetland Islands. We were to join *Cirolana* at Stornoway in the Isle of Lewis, not the easiest place to get to from the easternmost town in England. It involved three train journeys, the first to Norwich across the flat Broadland landscape of windmills and reeds, the second to the dingy recesses of Liverpool Street Station and the third, on the grandly named Royal Highlander express, from Euston to Inverness. The Royal Highlander was a night train and at that time the sleeping compartments were of the more spacious old type which included a pivoted device below the sink that bore the cryptic legend, 'This cabinet is not for solid matter'. I have always slept well in overnight trains, perhaps because the compartments are so like ship's cabins and the rumbling vibration is reminiscent of the ever-present sound of the engine-room. Having fallen asleep among the northern heights of London, I awoke just as the Royal Highlander was pulling out of Blair Atholl station in Perthshire, the two diesel locomotives straining on the first part of the long drag to Drumochter Summit, over 1,500 feet above sea level and the highest point on Britain's railways. I pulled up the blind just in time to see three black grouse, two blackcock and a grey hen, fly effortlessly over the train, their measured wing-beats belying a speed which easily outstrips the more common red grouse of Glorious Twelfth fame.

From Inverness we flew by Andover aircraft to Stornoway, a place of strange contrasts where Gaelic-speaking Pakistanis prosper during the week, the natives imbibe freely on Saturday night and the various species of free and not so free Presbyterianism reassert their iron grip on the Lord's Day. *Cirolana* lay alongside, her long fo'c'sle giving her a lithe appearance, most unusual in a stern trawler. I shared a cabin with David Ellett, an authority on the Wyville-Thomson Ridge, a submarine feature of the shelf-edge which marks the southern end of the Faroe–Shetland Channel. North of it, the bottom water is cold and Arctic in origin. South of it, in the deep recess of the Rockall Trough, warm Atlantic water supports a rich variety of fishes. The cabin we shared was above the propeller shaft, humming below us with all the smooth, under-stated power of a pre-war Rolls-Royce.

As the bows of this real lady of a ship hissed at over sixteen knots through the long Atlantic swell, we would not have swapped places with the Emperor of China himself. Our cruise was to take us the full length of the shelf-edge from west of the Shetlands in the north to the hake ground off south-west Ireland. In the process, *Cirolana* would steam through ten degrees of latitude. We knew that we could fish effectively with the gear we had down to 300 fathoms, but how would we get on at over double that depth? How long would it take for the net to reach the bottom? How often would we snag or even destroy the gear on any fasteners? Would we catch enough fish? Would the public be happy to eat them and, above all, to buy them? There were so many simple questions at the start, before we even began to think about the capacity of these special-ized fishes to go on supporting a fishery long after the accumulated 'standing crop' of sharks and bony fishes had been fished down.

Thanks to the skills of John Bridger and Schnozzle, we found that the gear worked well to at least 700 fathoms – over ten times the depth at which most commercial trawling is done – but we had to pay out all the heavy warp we had on each winch to get it down there. It took at least half an hour for the net to settle on the seabed and start fishing, but when it did, it worked better than we had expected. The enormous bobbins kept the foot-rope out of trouble most of the time. Every so often, the net would be torn in encounters with rocks and the mulberry-bush-like colonies of the cold-water coral, *Lophelia pertusa* (L.), but usually it came back more or less intact.

Although the gear was modified to cope with the special condi-tions, it was made up of familiar components and handled using Schnozzle's traditional expertise. What tumbled out of the cod end on to *Cirolana*'s deck after a deep haul was anything but familiar. The silver and soft browns and greens of the shelf fishes were here exchanged for a very different palette. The darkest browns and blacks were represented by sharks like *Dalatias licha* (Bonnaterre), known to trawlermen as 'Darkie Charlies'. Two sorts of smooth-head, floppy eaters of jellyfish called *Alepocephalus bairdii* Good & Bean and *Alecephalus rostratus* Risso, supplied the dark greys and fierce-toothed black scabbard fish, *Aphanopus carbo* Lowe, con-tributed the metallic glint of blued steel. So delicate was their skin that often the action of the trawl had removed it altogether to reveal the hard pink muscle below. The matt blue-grey of the round-nose grenadier fish, *Coryphenoides rupestris* Gunnerus, spilled over the sharks, their great eyes staring out of softly curved heads smoothly joined to bodies that tapered away not to a tail but a long trailing filament.

But if browns and greys were popular with the shelf-edge artist, they were not his favourite colours. He saved that honour for vermilion, expressed at its most brilliant in the orange roughys, *Hoplostethus atlanticus* Collett, deep-water red fish, *Sebastes mentella* Travin, blue mouths, *Helicolenus dactylopterus* (Delaroche), and countless deep-water prawns. He also liked a flash of silver, especially as represented by the greater silver smelt, *Argentina silus* Ascanius, distant relatives of the salmon family that, having been caught as the net was hauled, were often the only fishes on deck that retained a last feeble flap of life. Only the eyeless hagfish, *Myxine glutinosa* L., like lampreys ancient, jawless relics of a group which pre-dated the evolution of true fishes, had the power to retain their full squirming vigour at the end of their long journey to the surface.

As a naturalist whose previous experience had been restricted to the fishes of the continental shelf, I was fascinated, and the cornucopia of apparent rarities that poured out of Schnozzle's net was wealth indeed. If my first instinct was to study and dissect, my first duty was to help determine whether the fish resources of the slope were both palatable and substantial enough to sustain a fishery for the large British vessels which were to be displaced from the distant-water grounds. Ship's cooks are busy people, doing a thankless job in a topsy-turvy environment which would send the average celebrity chef shrieking for counselling. My arrival in the galley, carrying a fish box containing what the deck-hands had already learned from the tabloid press to call bug-eyed monsters, was as welcome as a pork chop in a synagogue. Particular exception was taken to the black scabbard fish, which were immediately condemned as 'long, black boogers with wicked fookin' teeth'. Remonstrating that Portuguese line fishermen prized them highly only made matters worse: I was adding a swarthy foreign connotation to their already unprepossessing appearance.

Somewhat abashed, I retired from the field and returned later with the fish reduced to innocuous-looking fillets, bereft of the offending heads, teeth, rough skin and other reminders of their alien provenance. This time I was not thrown out and I begged some roasting tins to see what the fillets tasted like. I was given a couple of shallow trays on the understanding that they would be put over the side once my cookery class was finished. A brush of oil, a few pinches of salt and examples of the least repugnant-looking fillets went under the grill.

John Bridger, who shared my enthusiasm for gastronomic adventure, and other nameless scientists who did not were by this time assembled in the saloon enjoying their nine-shilling-a-bottle gin. I emerged with steaming plates of orange roughy, black scabbard fish, round-nose grenadier and shark. Maybe the gin helped, but the orange roughy and the black scabbard fish were pronounced excellent. The grenadier was said to resemble whiting or saithe and the shark, the grey spiny species called *Deania calcea* Lowe, won an honourable mention as 'just as good as spurdog', a favourite with fish friers.

So far so good. Recalling our Lord's first miracle at Cana in Galilee, I had saved the best until the last. Of all the fillets, none looked whiter or freer of bones than those of the smoothhead, *Alepocephalus bairdii* Good & Bean. I had filled a whole tray with them and, with the accolades of my peers fresh in my mind, returned to the galley to fetch my *pièce de résistance*. I pulled the tray out from under the grill. To my surprise, the fillets no longer filled the tray but occupied about half of it and lay suppurating in a bubbling grey mother liquor which had appeared from nowhere.

Given that we were now steaming at full speed and that the ship was rising and falling sufficiently to test the stomachs of the weaker brethren, I thought it best to drain off the liquor and present the fillets on individual plates. Had I sampled the smoothhead first, my reputation as a fish cook might have been preserved and no one would ever have heard of the piscine delights of Padstow. One forkful was enough. The gelatinous flesh fell apart and a strange sneaky smell invaded the nostrils. 'Not much flavour but what there is, is nasty' was the kindest of the comments. The tray and the plates all went over the side at the insistence of the cook, and I never went near the galley for the rest of the trip.

By some miraculous and corrupt process, perhaps involving a case of beer and the good offices of one of the galley orderlies, John Bridger arranged a second, secret bite at the cherry in which only the two of us took part. We were both keen to try the greater silver smelt, *Argentina silus* Ascanius, fried briskly in the beef dripping the galley used to prepare their fish and chips. The results were so good that we had little room left for the final clandestine *bonne bouchée*, cutlets of the rabbit fish, *Chimaera monstrosa* L., with its close relative, *Hydrolagus mirabilis* Collett, easily the ugliest creatures to come out of the cod end. Their horrible bitterness also qualified them for being the most disgusting.

As the cruise progressed, it was clear from our results that, pro-vided there was a market for the sharks, the catch-rates of palatable species on the slope were probably high enough to support a viable fishery. Whether such a fishery would be sustainable, after the initial standing stock of older fish had been thinned out, remained doubtful. The immediate problem for the industry would be to convince the fish-buying public that some of these monsters tasted as good as the varieties they were used to seeing on the fishmonger's slab. Marketing and market research is a special-ized field we were glad to leave to others, like the White Fish Authority, who knew about such things. As naturalists, we could nevertheless provide a little background guidance on the biological basis for the differences in the visual attractiveness and palatability of the slope fishes.

So far as colour was concerned, the blue-greys, light browns and silvers adopted by fish living at the top of the slope or migrating up and down it were not so different from the colours of familiar fishes. The greys and browns provided good camouflage for fish living on the seabed and bright silver did the same for those living in mid-water. Further down the slope, where light levels are lower, darker browns and black offer one route to concealment, but it is not the only one. As depth increases, so both the quality and quan-tity of ambient light alter. Red light is least able to penetrate the deep sea and so a red fish or prawn at depth appears black. Fish like orange roughy are therefore able to use red pigments derived directly from their food to make them appear black without having to waste energy in the manufacture of the black-brown pigment, melanin. Inadvertently, by so doing, they make them-selves more attractive to the housewife.

The larger eyes of the slope fishes are needed to make the most of what light there is. They are not in themselves off-putting and, in any case, it is a problem easily solved: all the fishmonger has to do is cut off the heads. Odd shapes are another matter. One of the reasons that grenadiers look so peculiar is that they do not have proper tails but taper to a fine point. By dispensing with a conventional tail, they sacrifice propulsive efficiency but gain in a quite different way. When there is hardly enough light to see by, other senses increase in relative importance. Being able to detect objects through a sense of touch at a distance is a case in point. Many of the organs that confer this sense in fishes are borne on the so-called lateral line, a thin tunnel with openings along its length that extends down the middle of each side of the fish to the root of the tail. Touch at a distance is a short-range sense. Dispensing with a tail and continuing the lateral line backwards as a thin extension of the body increases its capacity to perceive and locate potential food, hard objects and threats. The result may look very odd indeed – grenadiers are probably best marketed as fillets with no bug eyes or rat-like tail to discourage the faint-hearted – but makes excellent biological sense.

The differences in the fishes' palatability were also explicable in terms of their natural history. The strong flavours of some sharks, taken to revoltingly unacceptable levels in the distantly related rabbit fishes, are caused largely by the high levels of nitrogen compounds in their body fluids, which do not have as much salt dissolved in them as the sea water outside. The nitrogen compounds help to restore the osmotic balance and thereby save the energy that bony fishes use up in swallowing sea water and excreting the salt. Palatability is not just about taste; texture matters as well. For fishes living on the slope, and especially those that migrate up and down it, the texture of the flesh and the control of buoyancy are linked.

Many bony fishes achieve near-neutral buoyancy, and thereby save energy, by controlling the amount of gas in an internal swim bladder. As the fish ascends in the water column, so the gas pressure in the swim bladder falls with the hydrostatic pressure. Left to its own devices, the swim bladder would expand and neutral buoyancy would be lost. Fish cope with the problem either by releasing gas through a duct (the so-called physostomous fishes) or by resorbing it through specially modified parts of the swim-bladder wall (the so-called physoclistous fishes in which there is no duct to connect the swim bladder to the outside).

One difficulty for both physostomous and physoclistous fishes is coping with gas leaks as they go deeper. Swim-bladder walls are flexible, just like balloons, and the pressure within them mirrors the pressure in the sea outside. As the fish goes deeper and the pressure gets higher, so a given quantity of gas occupies less space and provides correspondingly less buoyancy. Fish compensate by forcing more gas into the swim bladder to reinflate it and restore buoyancy. As the fish goes deeper still, so the gap between the hydrostatic pressure of the gases in the swim bladder and the partial pressure of the gases dissolved in the tissues of the fish, and in the sea outside, becomes progressively greater.

Keeping the wall of the swim bladder gas-tight is solved in fishes like the greater silver smelt by lining it with flat crystals of guanine, the same silvery compound that gives their scales their lustre and the fish its name. Thus, the greater silver smelt is able to buoy up much denser muscle and connective tissue than the distantly related smoothheads. Lacking the smelt's adaptations, smoothheads rely for their buoyancy on a high water content and make greater use of the low-density ammonium ion in their body fluids. As a result, the greater silver smelt has a delicious, well-textured flesh and the smoothhead is gelatinous and tastes sneakily of ammonia.

Grenadiers are also able to retain gas in their swim bladders at depth and can therefore sustain a relatively dense body which is palatable when cooked. The interesting exception was the orange roughy. It, too, has a swim bladder but has given up the struggle of trying to maintain buoyancy over a range of depths by alternately secreting and resorbing gas. Its swim bladder is filled with white fatty material and, as a consequence, has to be very much larger to provide the same degree of buoyancy. A large quantity of similar material helps buoy up its large head. Once again, excellent flesh quality is the result, but there is a penalty with direct practical fishery implications. Building up large quantities of fatty tissue is energetically expensive and slows down both growth and the time taken to reach sexual maturity. Stocks of slow-growing fishes are notoriously easy to over-fish, so orange roughy could never substitute for such fast growers as Icelandic cod.

As for sharks and rabbit fishes, they also use fatty material, in their case in their livers, to provide static lift and help counteract the weight of strong muscles and therefore firm flesh. The snag is that they also use nitrogen compounds to help balance the concentration of their body fluids with that of the sea water outside. Within limits, the taste of such fishes is still acceptable. But we knew, to our cost, that the flavour of rabbit fishes lay well outside these limits.

Had we found a viable alternative fishing ground for the British distant-water fleet? The answer was an emphatic 'no'. What we had found was a limited resource which, if carefully managed, could sustain a moderate level of fishing. In the event it was to be foreign vessels, especially from France and Spain with their more adventurous home markets, which would be the main beneficiaries. A few British and Irish vessels now take part and only time will tell how long the total fishing effort can be sustained. The possibility that the British distant-water fishing fleet might find economically attractive alternatives to their lost traditional grounds off Iceland and in the Barents Sea by fishing off West Africa and on the Patagonian Shelf in the South Atlantic also proved illusory. *Cirolana* never had the opportunity to benefit from her all-white livery, nor her capacity to distil her own supply of fresh water from the sea around her. Not long afterwards, and to the great relief of her Master, who hated to see the slightest rust streak against the white paint, her hull reverted to the traditional black of her predecessors. If anything, this change in livery enhanced her fine lines and her long fo'c'sle soon became a familiar sight in all of the home-water sea areas whose names are familiar to listeners to the BBC shipping forecast.

Not long after *Cirolana*'s repaint, the post I held was transferred to the Scottish Office's Marine Laboratory in Aberdeen. It was time for me to return to Scotland.

FIT LIKE?

Those whose knowledge of Scotland is derived from television plays and soap operas could be forgiven for thinking that it was rather a homogeneous place, peopled by a mildly prickly race who speak with Glasgow accents. The contrasts between the Ayrshire farmers of the Vale of Aylesbury, the couthy folk of Fife's East Neuk and the fishing communities of Angus and the Mearns had been my first introduction to the diversity of Scots life. It nevertheless came as quite a shock to find myself in rural Aberdeenshire among folk who substituted double 'e' for double 'o', 'f' for 'wh' and whose normal greeting was 'Fit like?' We had bought the partially converted remains of the third Seceder Kirk in Scotland in the tiny hamlet of Craigdam, seventeen miles north-west of Aberdeen.

The secession which led, in 1758, to the formation of the Seceder Kirk, was the result of a complicated series of squabbles which included groups calling themselves the Auld Lichts and New Lichts, the Auld Burghers and New Burghers. My friend from St Andrews days, the Reverend Ronald Ferguson, recently retired Minister of Kirkwall Cathedral in Orkney, once memorably classified the whole quarrelsome lot of them as 'Silly Burghers'. The particular issue which united the Seceders was their resentment over the influence which the local laird had in appointing parish ministers. They decided to set up their own Kirk, leaving the selection of the minister to the will of Almighty God, as interpreted by their own prejudices.

Finding a minister was one thing. Building a church to put him in was quite another. At Craigdam, the worldly but indulgent laird was a member of the great Gordon family, a distant ancestor of the recently deceased Marquess of Aberdeen and Temair whose highly entertaining directory of brothels briefly made him the darling of the tabloid press. Yet by eschewing lordly patronage, the Seceders had also cut off the help which landowning patrons normally gave in erecting and maintaining church buildings and paying ministerial stipends. So generous could this help be that a copper collection from the congregation each Sunday was usually enough to keep the finances in the black. At Craigdam, the financial circle was squared by making it a 'siller [silver] kirk' in which nothing less than a shilling was deemed an acceptable contribution to the Sabbath collection. So strong was local fervour that this tradition continued long after the Seceders and other free thinkers had rejoined the Church of Scotland.

*My sons Neil (left) and John (right) at
our house in Craigdam, Aberdeenshire*

Craigdam kirk was reached from the nearest village, Tarves (pronounced 'Tarvis'), by climbing a mile-long hill colloquially known as Futrat (ferret) Brae. Sometimes in the winter, the brae was blocked by snow and so, to avoid breaking the Sabbath, it was either Tarves kirk (a copper kirk) or burn with the rest of the damned in hell. On one such occasion, a local farmer came across a dear old wifie fully kitted out in her Sunday best but unable to get up the brae to Craigdam. He offered her a lift to Tarves which she accepted, on condition that she be allowed 'tae ging [go] hame tae change ma collection'. Craigdam rejoined the mainstream between the wars, but its underlying current of austerity remained until the kirk was finally closed in 1958. It was a distinction not always valued by members of the endowed congregation in the village which enjoyed the support of Lord Aberdeen and his more prosperous tenants. In the words of William Cook, farmer and antiquarian of Little Meldrum, 'You Craigdam folk are nae better than us Tarves anes, just mair fite-washened [white-washed]'.

Our immediate neighbours in the manse were Alfred Alexander, a retired farmer from Sanday in the Orkney Islands, his second wife, who was from Deeside, and her vivacious mother of ninety-six. Alfred was in his mid-eighties and, according to his wife, who was a little younger, his ambition, 'wis tae dee leavin' mair siller than his brither'. Her own ambition was to persuade him to let her knit on the Sabbath and to give up the awesomely strong Warhorse brand of pipe tobacco, the fumes from which permeated the entire house. You could not really say that the pursuit of his ambition had led Alfred into miserly habits. However, he begrudged matches when he could easily light his short Stonehaven briar with a rolled-up strip of the *Ellon Squeak* (the local paper, properly called the *Ellon Advertiser*). He also preferred simple local foods like kail broth and hairy tatties, the latter an amalgam of the rehydrated flesh of salt ling and mashed potatoes. As to drink, he liked nothing better than a fifty-fifty mixture of Bell's whisky and milk, the bottles side by side on the table. I never found out whether Alfred's fortune overtook that of his brother or not, only that Alfred's life was brought to a premature end when, not wishing to waste electricity by putting on the landing light, he missed his footing and broke his hip. He lingered on in the hospital, in his wife's words, 'Fair scunnert at nae gettin' a richt smoke', until at last his spirit 'fleggit awa'" to join those of his ancestors across the Pentland Firth.

The second Sunday after our arrival at Craigdam found the garden where the kirk had once stood echoing with metrical psalms. These are ordinary psalms tortuously rearranged so that they can be sung like hymns. They have long been popular in the Church of Scotland. The only snag is that the language of the King James Bible can sometimes end up like the conversation of Dennis the Dachshund (for those old enough to remember his starring role in the *Toytown* plays, one of the key ingredients of *Children's Hour* on the BBC Home Service). Indeed, an English cleric once complained that, when singing metrical psalms, he was never entirely sure whether he was praising God or God was praising him. Either way, the sound of the unaccompanied singing, in the still air of late summer, lifted the hearts of the Craigdam folk at their annual Service of Commemoration, and even the wasps behaved themselves. For us, of course, this opportunity to meet our neighbours so soon and under such happy circumstances could not have been more welcome.

Craigdam, during our time there, was less a recognizable hamlet and more a convenient way for the post office to find the scattered farms and cottar houses that straggle along the narrow, winding road leading out of Tarves to Old Meldrum. Most of the properties had at one time been part of the great Aberdeen estate. Roofed with slate from Ballachulish, and built of the subtly varied local stone, they grew out of the landscape as if placed there by the hand of the Creator himself. In fact, the hands that had put up the houses had also created the landscape. The legacy of the glaciers' retreat had been a generous crop of ice-worn rocks, from handy cobbles to slabs weighing several tons. Here and there, the neolithic farmers, who had walked the windswept slopes of north-east Scotland some 4,000 years before, had used the larger slabs to form their enigmatic stone circles. A handful of snow-white quartzite pebbles was thrown into each of the pits into which the upright stones were levered and, on the south-west side, a great recumbent stone waited to honour the full moon. The gathering-up of the smaller stones had to wait until the eighteenth and nineteenth centuries, when the great agricultural revolution, which had begun far to the south, finally worked its magic on the whinny knowes of the Aberdeen estate. To the iron men who cleared the ground and made the farms, the back-breaking toil would have seemed anything but magic. Perhaps, though, even they would have acknowledged that more than a little sleight of hand lay behind the mysterious process by which a heap of rounded irregular stones became a stock-proof dyke.

Thanks to the generosity of my farming neighbours, the dyke-girt parks (fields) that led down to the banks of the Youlie Burn were to be the scene of many a walk with the gun. Bags were small, but the burn would often produce a duck or snipe, the stubbles a brown hare and Crosshill's 'three-neukit parkie' (a three-cornered field) could usually be relied upon for a grey partridge or even that greatest of the rough shooter's trophies, a cock pheasant. They were happy days during which I also set in motion the shooting career of the son of a local farmer who had tragically died before he had had time to enjoy the privilege himself, and saw my own boys follow in the same footsteps using the tiny, hammerless .410 shotgun, proudly labelled Army & Navy (after the London store of that name), with which I had shot my first pheasant many years before.

Not long before our arrival in Craigdam, a spill of 'silage bree' (silage liquor) had starved the waters of the Youlie Burn of oxygen, killing many of its trout. It took over three years for the population to recover, by which time both my boys had learned to fish in an even smaller burn which flowed through Boghouse and other farms on the other side of Tarves. Both burns were tributaries of the River Ythan, famous among anglers for the quality of its sea trout. The distant ancestors of the Youlie's trout had also colonized it from the sea, but they had been isolated from it for many years by the lakes at Haddo House, the William Adam masterpiece that was the seat of the Gordons of Aberdeen. Spawning in the fords where the gravel had been washed fairly clear of silt by spates, the Youlie trout spent all of their lives below its overgrown banks, benefiting as much from the flies and beetles blown in off the fields as from the snails and insect larvae that shared their world.

Provided the fisher remains out of sight, brown trout can be as ardent in their pursuit of a bait as that favourite of the young angler, the perch, *Perca fluviatilis* L. For that reason, worm fishing tends to be looked down upon by superior persons and other miserable individuals who have long forgotten their youth. For boys learning about trout for the first time, days spent exploring small streams with the worm, a couple of split shot to take the bait down to the fish and nothing else (floats, bubble or otherwise, can be great frighteners of trout) teach lessons that last a lifetime. Often John and Neil would come back with two or three trout each, anything over four ounces counting as keepable and anything over half a pound a specimen to be whooped over. They learned to clean their catch at once (to prevent parasites in the gut from invading the flesh) by cutting through the pelvic girdle and extending the cut forward to the head and backward to the vent. A downward stroke with the edge of a spoon or the thumbnail to remove the kidney tissue that most anglers misidentify as congealed blood and the trout is ready for the pan. Even in death, wild brown trout are the most beautiful of fishes. They are wondrously spotted with vermilion and darkest brown against a ground of green and gold. How easy it must once have been to believe that all of this richness had been placed there by the Creator to delight the eye of Man! To remove the head before cooking spoils the trout's appearance at table and wastes the delicious meat in their cheeks. The boys invariably cooked theirs over an open fire, the skin crisply golden and the delicate pale flesh subtly scented with wood smoke. Not a scrap was left on the wiry little backbones.

The Youlie Burn is not the Thames and Craigdam is not Hurley, but John and Neil were now at the same stage in their angling careers as my brother Peter and I had been thirty years before, and their desire to catch the most fish was soon to be rewarded. It had rained heavily all Thursday and Friday, water splattering out of the rhones (gutterings), pouring down the road and spouting from the flower-pot-red land drains into the Youlie Burn. It was still raining on the Saturday morning, but at last the sun came out, steam rose from what passed for a lawn and the boys were digging furiously to fill their bait cans. The best pools in the burn extended downstream through a small plantation known as the Blairdy Wid (wood) and for half a mile or so as far as a farm called Backhill of Braiklay. A headlong rush to the top of the Blairdy beat found the Youlie Burn in furious spate, the water mud-coloured and the partly submerged stems of meadowsweet pulsing back and forth in the current. Extra split shot on both casts and the worms were tumbling about among the stems. First one rod was fiercely yanked and then the other.

Thank God for the six-pound nylon. Almost every cast produced a bite, the worm sometimes torn roughly clear to leave bare hooks but no trout. By the end of the afternoon, twenty trout, the largest an unheard-of ten ounces, lay on the bank. All the neighbours benefited, a family record was set and, by the following year, the boys were learning to cast a fly.

The Tolley ten-bore had not burned any powder since the great punt gun had put up the Blackwater widgeon. It nevertheless occupied an honoured place in the gunroom and, every time the game guns were cleaned, the barrels of the Tolley were pushed through and oiled and the outside wiped over with a waxed cloth. Even then it was nearly a century old but it wore its years lightly. Like many duck guns, it was a gun for special occasions and had not seen as much service as many a game gun a fraction of its age. The dry gunrooms and the care of my father had seen to it that it was still a credit to its long-deceased makers.

For such an old gun, the Tolley's appearance was surprisingly modern. At the time it was made, the boxlock action around which it was built, and which is still a bestseller among the guns of today, had not long been invented by William Anson with the help of his well-heeled colleague John Deeley. (Deeley is similarly associated with the well-known method of securing the forend to the barrels invented by John Edge, and this was also a feature of the sturdy Tolley.) For all its up-to-date lines, two features marked out the ten-bore as a member of the old brigade: it was designed and proved for use with the black powder that had brought victory at both Trafalgar and Waterloo, and the barrels were made not of steel but of three-iron Damascus. Damascus barrels were made by hammering twisted bars of wrought iron and steel round a mandrel, smoothing them off and boring them to create barrels of surprising strength. The convoluted patterns created by the mixed metals gave the barrels a functional beauty lacking in the pure steels now used in gun manufacture. They recalled similar patterns seen in the Damascus swords which were prized by Saladin's warriors for their ability to hold an edge but not to snap in action, and rightly feared by the Crusaders, who often had to make do with softer-edged weapons made from an early type of wrought steel.

That morning in late November, on the banks of the Ythan Estuary, there was just enough light to make out the patterns in the thirty-two-inch tubes when the pinkfoot geese began to speak. Out on the tideway, their wild chorus rose and fell, and each time it rose, their music grew louder. My supply of three-and-a-quarter-inch cartridges for the Tolley was nearly at an end. The cases were no longer made and even the best of my home-loaded remnants were looking distinctly dog-eared. I had loaded them with the last of my black powder, topped with compressed newspaper and BB shot left over from my father's wildfowling days. My hope was that the forgiving black powder would make up for the deficiency of the wadding and the weak turnover of the veteran cartridge cases.

Another crescendo from the restive geese and the first party is on the wing, hounds of heaven indeed but nowhere near my hiding place among the snow-sprinkled whin bushes. Another check to make certain that no snow has found its way into the muzzles and I slip two of the least tatty-looking cartridges into the breech. Bitter cold steals the strength from the strongest of fingers and, leaving my left hand to fend for itself and look after the Tolley, I warm my right hand under my armpit so that, should the need arise, I can still operate the safety-catch. Another flapping cacophony out on the flooding tide and a large party takes off, once again on a bearing that will not take them close enough to me for the overhead shot which alone, on such a still morning, would stand a chance of bringing a grey goose humanely to bag.

By this time my left hand is almost numb despite its shooting mitten and, holding the heavy Tolley round the small of the stock with my now functional right hand, I attempt to revive the left by blowing on it, the resulting condensation, if anything, making it feel colder. For a moment or two, I forget about the geese in my efforts to restore feeling to my hands. 'Wink-wink, wink-wink' – I look up to see two almost overhead. There is no time to stop the gun as the long brown barrels swing beyond the leader's head. A gout of orange flame, a cloud of grey smoke and burning fragments of the *Daily Mail* among the whins silence the remaining legions out in the estuary. High above, a single goose folds and begins its long descent to fall stone dead behind me. It is a moment of triumph, and any regret I feel is tempered by the thought that my goose will feed six to the battery chicken's two and, while the adult pinkfoot has known nothing but years of freedom, the chicken has been a prisoner for all of its short life.

Back at the house, I wash the black powder fouling from the right barrel with boiling water and so, in the language of the old 'fowlers, sweeten my gun. I make up a cloudy mixture of Young's .303 cleaner and water to remove the last traces of fouling and permeate the kitchen with that wonderful coal tar smell so dear to old sportsmen. Dried and oiled, the ten-bore's barrels are left overnight in the warm kitchen, dried and oiled again in the morning and reunited with the action to resume their long sleep, muzzles downward, in the gunroom.

AFTER MᶜINTOSH

The Marine Laboratory in Aberdeen which I had joined had its complex origins in the tortuously Scottish processes which were also to starve the great Professor McIntosh of St Andrews of funds, at the very time he was revolutionizing our understanding of the life cycles of the main British food fishes. The Fishery Board for Scotland had convinced itself that the natural production of young fish in the North Sea could be materially increased by the release of fry reared in a hatchery. By 1892, a hatchery and laboratory were established at Dunbar in south-east Scotland and, over a period of six years, some 152,000,000 fry, mainly of plaice, had been released. The Dunbar site lacked a sea-water enclosure large enough to hold adequate numbers of adult brood fish and so, at the end of the 1899 rearing season, the whole complex was dismantled and taken to a site at the Bay of Nigg near Aberdeen where a laboratory had been erected the year before.

The release of reared plaice fry from the Bay of Nigg ceased in the early 1920s without demonstrating any benefit to the fishery. Field experiments, especially on a scale large enough to perturb marine populations, are notoriously difficult to plan. Plaice are highly fecund fishes and the numbers of fry produced by the Nigg and other European hatcheries would have been overwhelmed by the results of wild spawning. Given year-to-year variation in the survival of the fry, and the fact that the wild spawning stock had increased because of reduced fishing during the First World War, there was never a chance that the Nigg Bay experiment could produce a detectable result within any reasonable period of time. This outcome could have been predicted by a careful comparison of McIntosh's work on fecundity with the statistics of plaice catches, but it was to be some time before the necessary synthesis of biology and mathematics developed. When it did, it also became clear that, provided there were enough spawning adults in the first place – and there certainly were in the 1920s – the numbers of fully formed young plaice 'recruiting' to the fishery were limited, not by the numbers of fry *per se*, but by the quality of their early environment.

McIntosh's descriptive accounts of successive life cycles laid the foundation for such insights, but it was to require the mathematics of the actuary to lay them bare. A wag once described an actuary as an individual who found accountancy too exciting. Actuaries earn their money by working out the average chances of people dying over a range of ages. Constructing such life tables for long-lived mammals such as ourselves is indeed a slow-moving pastime. By comparison, the early life-stages of fishes pass with electrifying speed.

One of the first scholars to recognize that the lives of both fish and men could be analysed by actuarial techniques was D'Arcy Wentworth Thompson, the Anglo-Irish Professor of Natural History at University College, Dundee, who was later to be McIntosh's successor at St Andrews. The characters of the two men were very different. The single-minded intensity of McIntosh was at its best in exact description through attention to the finest detail. Although not opposed to the admission of women to the university, McIntosh was most at ease in male company and, although something of a chatterbox himself, he nevertheless had little time for talkers who 'can do nothing else'. Mark the contrast with

D'Arcy Thompson, who was effortless superiority incarnate, at ease in any company and charming to ladies of all ages until the very end of his long life.

The son of the Professor of Greek at Galway, D'Arcy Thompson was as well-versed in the dead languages as he was in the natural history and mathematics of his day. After achieving the Presidency of both the Classical and Mathematical Associations, he then made his greatest contribution to the science of natural history with *Growth and Form*, an application of mathematical insights to the study of organic structure that was revolutionary in its day. It is written in a style so gracefully economical that, long after the obsolescence of its subject matter, it is still regarded as one of the finest pieces of scientific writing in the English language. It was fortunate, therefore, that such a man was available to adorn the Fishery Board for Scotland from 1898 and to help found the International Council for the Exploration of the Sea, to which he was appointed a British representative from 1902. Although D'Arcy Thompson's gifts to the cause of Scottish fishery science were mainly bestowed from the position of monitor rather than that of the working-deck

Professor D'Arcy Wentworth Thompson

participant, these gifts were to prove considerable. His emphasis on the value of keeping accurate statistics of the size and structure of catches and on using them, both as records of an important industry and potential indices of the stocks of fish from which they were obtained, was to be of lasting value to the cause of fishery research in Scotland.

I never met D'Arcy Thompson, but I knew many who had. They included my wife's uncle, the sculptor Alfred Forrest, whose accurate interpretation of D'Arcy's leonine head now graces the library of St Andrews University, and his talented former student colleague David Burt, who had fought with the Black Watch in the First World War. Burt recalled D'Arcy's advice on the preparation of lectures (the subject of D'Arcy's own first effort had been *Amoeba proteus*, at that time regarded as the most primitive form of animal life): 'After the first five minutes, I had told the class all I knew about *Amoeba proteus* and, by the end of the hour, all I knew about the entire animal kingdom.'

Once, chided for his kindness to undergraduates, especially attractive female ones, D'Arcy Thompson asserted: 'I see nothing wrong with spoon-feeding my students, but I have the strongest objection to working their mouths for them.' On another occasion, one kindly young lassie, filled with compassion at D'Arcy Thompson's momentary hesitation while reading an account of Mediterranean fishes in the classical period, sought to reassure her ancient professor by imploring him to take his time. 'Indeed I must, young lady,' he replied. 'This Greek text was written in the fourth century BC and I have come across a word which had several meanings, even twenty-four centuries ago.'

During his time in Dundee, D'Arcy Thompson was a leading

light in the Homeric Club. It was their custom to meet out of the city. One summer evening, they had gathered in the bar of a very traditional but respectable hotel in north Angus. A couthy local, no doubt in a convivial frame of mind, asked D'Arcy who he and his exotic-looking chums were. 'We are the Homeric Club of Dundee' was his response. The reply was equally to the point: 'Oh aye, whaur's yer doos [homing pigeons]?'

D'Arcy Thompson retired in 1948, his contribution to the story of fishery research in Scotland by then a distant memory.

In 1923, the Marine Laboratory was relocated to a site in what used to be the fishing village of Torry and is now a district – albeit a highly distinctive one, with its own traditions of 'spikkin' (speaking) and drinking – of the city of Aberdeen. For many years, the story of fishery research in Aberdeen was to be one of worthy progress. If Lowestoft was to take the lead in developing the 'theory of fishing' and techniques for assessing the status of fish stocks, Aberdeen was to make important contributions to our understanding of the eco-logical basis for fish production and of the mechanisms by which trawl and seine nets catch fish. By the time of my first visits there in the 1960s, the Marine Laboratory in Aberdeen had enjoyed some sixteen years of the direction of Dr Cyril Lucas. Although he was an accomplished scientist in his own right, with a special interest in the chemical defences of certain phytoplanktonic organisms, it was Lucas's skills as an organizer and team-builder which were to mark him out for high honours. Uniquely for a serving member of the scientific Civil Service, he was to be elected to Fellowship of the Royal Society, not for his personal contribution to science, but for his outstanding services to fishery administration.

Intelligent, incorruptible and absolutely to be relied upon in a crisis, senior civil servants of the old school are a dying breed. Understatement, in the interests of keeping problems as far as possible from ministers' desks, is one of their chief attributes, as Lucas once famously demonstrated. An inshore fishing vessel, on passage from the west to the east coast via the Caledonian Canal and Loch Ness, had reported the sonic imprint of the Loch Ness monster on its echo-sounder trace. The tabloids made their usual most of the report but, shortly after, assumed a more strident tone when, in response to their desire to publish the trace, they were told by the fishing vessel's skipper that the echo trace was now under lock and key in the Marine Laboratory, Aberdeen. Talk of a government cover-up sold even more newspapers and, not surprisingly, an irate letter demanding the truth landed on Lucas's desk. Chuckling quietly, he looked again at the skipper's echo trace, with its crude representation of a long, matchstick-thin beastie with four well-separated legs and a wavy tail, and formulated his reply:

> Dear Madam,
>
> Thank you for your letter asking about the recent report of the presence of a monster in Loch Ness by the skipper of a motor fishing vessel on passage via the loch. Doubts, which I share, have been expressed about the authenticity of this particular record.
>
> Yours sincerely,
>
> C. E. Lucas

No more was heard of the monster that summer or, for that matter, of the skipper or the irate correspondent.

Lucas was an adept fisher of men, and the appointment which gave him the greatest satisfaction was one of his earliest. He was well aware of the need to secure the services of a colleague equipped to apply the latest thinking on the dynamics of exploited fish populations to Scotland's fisheries. In Basil Parrish, the scion of a yeoman farming family from Cambridgeshire and a former RAF operational research colleague of the Lowestoft laboratory's director Michael Graham, he had found his man. Parrish was a master of his subject in its most up-to-date manifestations. He was also confident and gregarious, and his gifts lay less in innovation and more in putting the lessons of good science to work. In the latter, he was outstandingly successful, but the achievement of which he was most proud was his election to the membership of the Marylebone Cricket Club. Although an accomplished cricketer, Parrish was ever sensitive to the feelings of others, and only occasionally did he inflict the MCC's appallingly garish tie on his colleagues.

By the time I joined Basil Parrish in Aberdeen to help him draw the various strands of British fishery research together, integration of the English and Scottish programmes had been taken as far as the rival Fishery Departments were prepared to allow. This was disappointing, of course, but it left me with enough free time to pursue my long-standing interest in the living relatives of the earliest fish-like vertebrates, the lampreys and hagfish.

THE PRESENT PAST

When we were children, the annual visits to our beloved Norfolk coast never lasted long enough. All too soon it was time to pack the tents away and start the journey home in the Ruby saloon. It was a long goodbye; the little Austin was not happy above 35 mph and, for a time at least, the sandy fields of north Norfolk reminded us that it was not so long ago that they had felt the surging tides of the North Sea. Back then, in those distant days long before the Kaiser's War, it had been called the German Ocean. Ten miles inland, the strangely haunted village of Walsingham, with its creepy shrines and idolatrous processions, still retained in its winding streets a little of a seaport's atmosphere. Once it lay behind us, only the pantiles on the flint-faced cottages – brought over as ballast in Dutch merchantmen – preserved a last link with the coast we were so loath to leave. Soon even they were gone and, eventually, we would find ourselves back in Buckinghamshire, the grim prospect of the long autumn term looming ever closer. The trophies would be unpacked the following morning, temporarily banishing such melancholy thoughts, and placed in the corner of the bedroom we had set aside as a museum. Shells of horse mussels, red whelks and hard shell clams from Holkham, oyster and cockle shells and the cast skeletons of shore crabs from Wells – all took their place alongside the blad derwrack still attached to its flint pebble. If the strong-smelling bladderwrack was soon to be banished to the garden shed, the other specimens would remain to remind us of past magic.

They were not to be our only reminders of the sea. The Chiltern

Hills that overlook the Vale of Aylesbury are made of chalk, compacted remnants of marine organisms that last basked in the seas of the Cretaceous era over sixty million years ago. Most of the chalk is made of such tiny creatures that they are not distinguishable without a microscope, even to the keen eyes of small boys. Among them, though, are larger fossils of sea urchins and internal moulds of large clams, the latter a little like the ones we had found at Holkham. Not all were so familiar, however, and especially was this true of the strange belemnites, siliceous bullet-like remnants of extinct relatives of squid and cuttlefish. How could it be that, all those years ago, the very middle of England lay under a shallow sea – a sea, moreover, that was host to a fauna more typical of warmer latitudes than of the windswept sand bars of north Norfolk?

Without knowing it, we had in our museum direct evidence that it is not only coastlines which change with time. Even whole continents are not fixed on the earth's surface but are carried as buoyant granitic passengers on a mobile basaltic sea floor, created and spreading at mid-ocean ridges and sinking into the molten mantle at deep ocean trenches. Together, the pieces of land and seabed are known as tectonic plates. They move slowly – the Atlantic, for instance, is getting wider at the rate of a thumbnail a year because of new crust being formed along the mid-Atlantic Ridge – but move they do and hence the Chilterns' warm Cretaceous sea.

Sixty million years is a relatively short time in the history of life on earth and that is why the fossils in the chalk looked familiar. The further back one goes, the less familiar the fossil fauna becomes. Nevertheless, there are some organisms that have remained relatively unchanged over several hundred million years. Ruthless as the effect of natural selection can be in eliminating organisms unable to cope with a changing environment, or with newly evolved competitors, it also provides ample evidence that, 'if it ain't broke, don't fix it'. Among fishes, two of the best examples are the eel-like lampreys and hagfishes, forms which go back to the very dawn of the evolution of fishes when, in some groups, anterior gill arches were evolving into the first vertebrate jaws. Critical differences in the relative positions of the gill skeleton and the gills themselves meant that the predecessors of lampreys and hagfishes were excluded from this revolutionary development. That they did not become extinct was due to their early success in adopting an unarmoured, eel-like body form to fit them both for burrowing at key times in their lives. Like most successful but archaic organisms – and we think that animals like them have been around for over 400 million years – species of lampreys and hagfish are found all over the world. To be able to make a contribution, however small, to our understanding of such living fossils is indeed a special privilege.

Many years before, in examining lampreys I had collected in the River Chess, I had noticed a column of tissue above the flattened nerve cord which extended for most of the length of the animal. Dissatisfied with sketchy representations in dissection guides which seemed not to distinguish this structure from muscle blocks, I had toyed with the idea that it might play a part in keeping lampreys the right way up by raising their centre of buoyancy above that of gravity. My later observations of whole dead lampreys shed doubt on that idea – doubt which was confirmed when I noticed that a dissected section of the obscure tissue sank rapidly. Clearly lampreys stay the right way up, not by any passive mechanism, but by responding actively to signals from the organs of balance in their heads. Relying on active stabilization in this way demands a lot of the central nervous system, but greatly improves manoeuvrability by reducing inertia. Many birds also depend upon active stabilization for the same reason, as does the latest generation of air-superiority fighter aircraft.

I duly got in touch with the great lamprey enthusiast, the late Lord Richard Percy of Newcastle University, and he generously gave me two valuable pieces of information. He had discovered that the tissue which had baffled me was the centre of red blood cell formation in lampreys. To my surprise, it was performing the function of bone marrow in an animal whose skeleton was of cartilage and so did not have a bone in its body. Lord Richard's second gift was to tell me that there was a large population of the hagfish, *Myxine glutinosa* L., in the muddy sediments of the Farn Deeps off the coast of north-east England.

An early representation of the hagfish or borer, Myxine glutinosa *L.*

Hagfish are creatures of which nightmares are made. Floppy, eel-like horrors about a foot in length and with no external eyes to light their path, the hagfish of European seas burrow in the soft muds found over large areas of the continental shelf at depths greater than fifty fathoms. Four barbules, exquisitely sensitive to the chemical and mechanical properties of their prey, project around their snouts like the petals of a flower which has somehow acquired the power to squirm. As to their colour, not for hagfish the smart slate above and pearl below of our Chess lampreys; somehow the uniform livery of the hags combines muddy grey with hearing-aid beige in a hue for which the English language has no exact name but which forensic pathologists would recognize as redolent of many a tented exhumation under the dripping yews. Like a lot of soft-skinned animals that burrow, hagfish depend upon slime to protect them from cuts and to carry the antibodies that defend them from bacterial attack. So slimy are they that a single individual was said to be capable of turning an entire bucket of sea water to jelly. And, like the pathologists, they were believed to make their living from carrion.

I had never believed the bucket story but I had no doubt that, given the opportunity, hagfish were partial to carrion. Indeed, as the Aberdeen laboratory's rare fish expert, I had once had the grisly duty of identifying the squirming forms of the hags clustered round the body of a drowned oil worker, God's broken image encased in neoprene: somehow the diver's scent must have oozed through a seam in his suit to excite the eager hags. However, depending upon carrion is a very uncertain way of life. On land, vultures succeed by climbing high in thermals and scanning large areas with some of the keenest eyes in the animal kingdom. How could hagfish hope to be as successful, bumbling about in the murk with only their chemical and tactile senses to guide them? There was only one way to find out: to ask the hagfish themselves.

As a result of my early interest in lobsters, I was already in close touch with the laboratory's shellfish research group and I knew that, every so often, they chartered small fishing vessels to sample pink shrimps, *Pandalus borealis* Kröyer, in the Farn Deeps. Yes, hagfish often came up in the shrimp trawl and yes, I would be welcome to join them on their next charter. I did not have long to wait. Gut-wrenching experiences aboard the *Tellina* in the same sea area had faded from my memory, much as women are alleged to forget the agonies of childbirth. The young skipper and crew of the aged but capable shrimp trawler could not have been more solicitous as we set off from 'canny' Shields on the long steam to the Farn Deeps. They offered me partially cooked fatty bacon, washed down with strong tea cooled with tinned Carnation milk. Such homely fare can be an anti-peristaltic combination to those of delicate, land-softened tastes. A few years before, such a mixture, offered in the diesel-fumed fug of a fishing boat's fo'c'sle, could have ended only one way. But on this occasion, I was so preoccupied with trying to make sense of my new friends' Geordie dialect that I defeated even the best efforts of my autonomic nervous system to reject the hospitality of their table.

As with so many fisheries for species that fetch a lot of money, getting enough of the pink shrimps usually involves catching a much larger volume of other creatures, mainly, in the Farn Deeps, whiting, *Merlangius merlangus* (L.), and haddock, *Melanogrammus aeglefinus* (L.), most of which are discarded as being too small. The shrimps were retained along with the hagfish, which were taken at the rate of one to four per half-hour haul. By the end of the week, we had sixty-five of them and not once had we seen the slightest evidence of their supposed skills in turning buckets of sea water to jelly. Back alongside, Dr Jack Buchanan, director of the Cullercoats Laboratory, donated another twenty-two specimens he had dredged directly from the mud of the Farn Deeps, and fishermen working the Fladen Ground in the north-central North Sea gave me another forty-two. In the laboratory, the collection told a story that belied the hag's macabre reputation.

Some of the specimens contained bitten-off pieces of flesh that probably indicated feeding on dead fish discarded from small trawlers like ours, but easily the commonest items were active animals like shrimps, small squid and worms that normally crawl over or burrow into marine muds. One of the hags even had the wing of a wasp and another a parasitic louse that had probably fallen off a preening sea bird. Most bizarrely of all, one of the Fladen hags contained the torn-open remains of several mature eggs of its own species. So far as we could tell, after looking at more hagfish stomachs than any of our colleagues, hags are determined predators that locate their prey above and within the surface of the sea and seize it with powerful bilaterally opposed nippers. Given access to carrion, they are attracted to it and readily bite into it, like many other marine animals. However, they are not dependent upon such free lunches but fill an ecological niche closely paralleling that of large ragworms, whose shape and bilateral mouth-parts they share. Uniquely among vertebrates, the total blood salt concentrations of hagfish also resemble those of marine worms in being close to those of the sea water around them. They fall way outside the evolutionary lines that led to the true fishes or indeed to ourselves. How astonishing, therefore, to consider that at some point, over half a billion years ago, we must have shared a common ancestor.

—

AMONG THE BLUE MEN

—

To William McIntosh, looking out from his laboratory across St Andrews Bay to the Bell Rock light and beyond, so vast was the sea

Mending lobster pots damaged by a storm

compared with the burns and ponds of his country childhood that its resources seemed inexhaustible, and counting them, species by species, the stuff of dreams. Even nowadays, in the light of many years' experience of applying the techniques of advanced fishery science, it still seems something of a miracle that scientists and fishermen can meet every year for well-informed debate on the state of the fish and shellfish stocks on which both depend for their livelihoods. It was McIntosh's fiercest critics – the fishermen who knew that, as year succeeded year, they were having to work harder to maintain their catches – who first pointed the way to the insight that catch per unit of fishing effort (CPUE) could provide an index of the state of the stocks which they were harvesting. It is a simple enough technique which, so long as the catchability of the target species is known, can, over a period, produce useful information on the relative abundance of the resource at the time the fishing took place. Records of CPUE, however well gathered and analysed, are nevertheless only indices. They do not, on their own, provide direct estimates of the size of fish and shellfish stocks. This greater task, with its infinite scope for argument among scientists, fishermen and politicians, is the preserve of the fish population dynamicist.

Lobsters are called 'blue men' by the fishermen of the far west of Scotland because their mottled blue-green livery is a reminder of the ancient Gaelic legend of the blue-green beings who are said to come dripping aboard, singing an elaborate song. If the awestruck fishers are able to continue in the same metre and rhyme, the blue men return over the side and disappear harmlessly below the waves. But if the song is interrupted, the blue children of the sea take the boat and its crew with them. It is a very old myth in which a hazy belief lives on, especially in the Catholic realms of Gaeldom. The modern link with lobsters has a certain rightness to it in that they also are ancient beings whose capacity to sustain artisanal fishing is as old as the Gaelic world itself.

For many years, John Stewart, lobster fisherman of Grimsay, North Uist, in the Outer Hebrides, had provided a meticulous account of his catches of the blue men and of the effort that had gone into obtaining them. Only the three-week gaps set aside for the Communion Season of his Gaelic-speaking fellow Presbyterians interrupted the graphs which, by the time I had the opportunity of inspecting them, were among the best records of a lobster fishery in the country. Elsewhere around the coast of Scotland, Tom Meldrum of St Andrews and other fishermen contributed their records – some, like John's, indicating high but declining catch-rates of large lobsters, others, like Tom's, poorer but stable catches of smaller ones. Although not large in terms of its landed weight, the Scottish fishery for lobsters had long been regarded as something of a flagship among the shellfish resources because of the high price each lobster could command. But there was increasing concern that the inshore populations on the east coast were already substantially over-fished and that the more recently established fishery to the west of the Outer Hebrides was in danger of going the same way.

How could we find out if these concerns were soundly based or merely the nostalgic reminiscences of old men? The stories told by the catch records were not enough on their own to guide the future management of the fisheries; we needed to know a lot more about how many lobsters were out there sheltering in the lee of the jagged rocks of St Andrews Bay and the Atlantic coast of the Western Isles. Our colleagues studying the related species *Nephrops norvegicus* (L.), the scampi of commerce, could estimate their abundance by counting the entrances of their burrows using a television camera suspended below a research ship. No such simple method was available to survey the stocks of lobsters, much rarer creatures that spend long periods concealed among rocks and kelp. No, we had to fall back on more indirect methods and, in outlining what we did, it is difficult to avoid the technical language that fishery scientists tend to use among themselves.

One of the ways in which we assess the status of fish stocks is to follow the fate of individual cohorts – fish born in the same year – over a period of years. Given an estimate of natural rates of loss and accurate catch statistics, this cohort analysis technique can be used to calculate the size of the stocks from which the catches were obtained. If we can also estimate the rates at which fish die as a result of fishing and natural processes and compare this estimate with their rate of growth, we can determine whether the current combination of fishing mortality rate and size at first capture is making the best use of the resource or leading to the deaths of too many fish before they have realized their growth potential. On its own, the cohort analysis technique can also be used to look at trends in the numbers of young fish entering the fishery each year and in the numbers of their adult parents. One way and another, cohort and other approaches to so-called virtual population analysis have a lot to recommend them for species which can readily be 'aged'. But as we have seen, lobsters, in common with other crustaceans, cannot be aged as fish normally can by examining ear bones or scales.

Fortunately for me, an outstanding Aberdeen-based fish population dynamicist, the late Rodney Jones, had realized that, although it was not possible to tell the age of an individual, it should be possible to estimate the age of lobsters within a given size-range provided the dynamics of their growth were known, and thence to subject them to a modified cohort analysis in which growth parameters substitute for ages. Only the lack of growth parameters for lobsters prevented the immediate application of Jones's technique. The rigid external skeleton of lobsters, which is so efficient mechanically and provides such robust protection against predators, has one great disadvantage. Growth requires the animal to moult and swell larger, whereupon, for a short time, the lobster is soft and vulnerable. To estimate growth parameters for lobsters, therefore, requires information both on how much the animal swells after a moult and how often moulting takes place. There is no short cut to obtaining the necessary information. It requires large numbers of lobsters to be tagged, recaptured and recorded in representative fisheries. We could do the tagging and analyse the results, but only the fishermen could supply the lobsters and record the recapture and release of tagged animals.

In John Stewart and Eric Twelves in the Western Isles, and Tom Meldrum in St Andrews Bay, we knew that we had men hauling the creels and keeping track of 'our' lobsters who were as committed to the work as ourselves. As a direct result of their diligence, we were able to demonstrate that, although the rate of fishing to the west of the Outer Hebrides was much lower than that in St Andrews Bay, the fishery would benefit in the long run from allowing the lobsters in the catch to grow larger, by fishing less intensively and not retaining the smallest specimens. In St Andrews Bay, so intensive was the fishery that the catches consisted almost entirely of small, young lobsters which had not had time to realize their growth potential or even, in many instances, to breed. Whereas the Western Isles fishery was a little wasteful in terms of yield per recruit, that in St Andrews Bay was grossly so and risked stock collapse from lack of the large lobsters required as the parents for future generations.

Although the analyses of the lobster fisheries of the Western Isles and of St Andrews Bay had been rewardingly clear-cut in indicating what needed to be done, they shed no light on the ecological background to the assessments, and their results raised both economic and biological questions. The Western Isles creel fishery was, and remains, economically viable on the strength of the lobster catches alone. That in St Andrews Bay is sustained economically by the catches of edible crabs, *Cancer pagurus* L., which are the main object of the fishery; the lobsters are merely a valuable bonus. So productive is the local crab stock that it can readily sustain a level of fishing effort which dangerously depletes the lobster stock. What was remarkable biologically was that the greatly depleted stock of lobsters showed no sign of being limited by lack of young, despite the fact that many of the adults were caught and landed before they had had a chance to breed.

In one important respect, the two fisheries were very different. Lumbering and heavy out of water, lobsters in their natural environment are as light on their feet as ballet dancers. They are easily displaced by the effects of strong tides and grounding waves and do best in the areas of lee around rocky outcrops where they live on the fauna inhabiting the surface of rocks and the top few centimetres of any sediments in their territory. So extensive is habitat of this kind among the outcrops of Lewisian gneiss to the west of the Outer Hebrides that large populations of lobsters can be sustained, as can a fishery which is both biologically and economically viable on the basis of the lobsters alone. By contrast, in St Andrews Bay the areas of lee between outcrops are restricted to the sublittoral zone, the rest of the bay having a sandy bed which is ideal habitat for the edible crab. Unlike lobsters, which are surface grazers, edible crabs dig deep into marine sediments, creating pits up to nine inches deep and thereby securing access to far greater food resources than are available to lobsters.

Fortunately, the situation of the lobster stock in St Andrews Bay was not as dire as that of the North Sea cod because of the special features of lobster reproduction. Lobster larvae hatch directly into the plankton from eggs held externally on the underside of the female adult's tail. There are four larval stages, which take some six weeks to complete. Experiments we undertook in deep plastic enclosures moored in Loch Tournaig in Wester Ross showed that the larvae tended to accumulate in the upper layers of the sea, the third and fourth stages being drawn right to the surface as neuston. Because waves driven by onshore winds inevitably have a longer fetch than waves driven by offshore winds, objects on the surface tend to be driven into the shallows. By making a short offshore migration prior to hatching their eggs, lobsters use this simple physical mechanism to ensure that, at the time their larvae are ready, after four moults, to seek a living on the seabed, they do so close to where their parents grew successfully to adulthood. Mercifully for the St Andrews Bay population, though, not all of the adult offshore migrants returned inshore. A few of the larger animals remained on reefs too far offshore for most of the local vessels to exploit and it was almost certainly their larvae that sustained the fishery. (How long this fragile source of young will remain is uncertain, however, given the current interest in exploiting the offshore lobsters using larger vessels and larger traps.)

The lobster, *Homarus americanus* Milne Edwards, that inhabits the eastern coasts of North America is very slightly different from our own, a little chunkier perhaps and with a brownish cast to the blued steel of its European counterpart. It is classified as a different species but, in all that matters, it is the same animal and sustains an inshore trap fishery similar, but on a larger scale, to those of northwest Europe. Unlike *H. gammarus*, the North American lobster also supports a deep-water fishery far offshore for large lobsters. These leviathans, tens of pounds in weight, are found among the submarine canyons off New England in hundreds of fathoms of water and they may well contribute larvae to the inshore stocks in the same way that, on a much smaller scale, the offshore reef lobsters apparently contributed recruits to St Andrews Bay. For many years, we had speculated that Rockall Bank and the continental slope to the west of Scotland might also harbour an important lobster resource. In our wilder imaginings, we had even speculated that a few *H. americanus* larvae might have been driven by westerly gales for 2,000 miles across the Atlantic to find a haven among the reefs surrounding the Rockall Islet, Scotland's lonely outpost 150 miles west of the Outer Hebrides. These tantalizing possibilities had not been explored: no ocean-going vessel could be spared from what were then regarded as more important surveys of finfish resources. We would have to wait – but not for long.

SOMETHING TO DO
WITH THE WEATHER

The separation that exists between the fishery administration of Scotland and that of the rest of the United Kingdom has never been other than a thoroughly bad thing, inimical to both healthy scientific debate and logistical efficiency. Just occasionally, though, the ridiculous arrangement would produce an unexpected benefit and of one such I was to be the grateful recipient.

Not to be outdone by the Ministry of Agriculture, Fisheries and Food's *Cirolana*, the Scottish Office put in hand what they hoped would be an even better vessel, *Scotia*, based on the same hull and built in the same Clydeside shipyard. Lacking a Gentleman Geoffrey, the Aberdeen laboratory fell back on that dreary offshoot of Presbyterian culture, decision-making by committee. If, in Charles I's words, Presbyterianism was 'no religion for a gentleman', it was also to prove that it was no religion for a naval architect.

Research vessels are like warships in that they are required to combine a range of ostensibly incompatible capabilities in a small package able to survive and function in the worst of seas. The needs of hydrographers, planktonologists, acoustic and gear experts, fish population dynamicists and benthic ecologists often differ widely. They had already been met in *Cirolana*, and a similar vessel, with a larger working deck to fulfil the special Scottish requirement to handle the largest commercial trawls, would have filled the bill admirably. Sadly, all that was retained of the previous successful design was the hull (but built a deck lower). *Cirolana*'s two in-line diesel engines, supported on a massive raft to reduce vibration, were replaced in *Scotia* by three enormous V8 engines mounted on a raft entirely inadequate to damp the mind-numbing vibration they generated. Where *Cirolana* cruised comfortably at fourteen knots and could make sixteen with the wicks turned up, poor *Scotia* normally lumbered along at eleven. The enlarged working deck of *Scotia* was well laid out and, unlike *Cirolana*'s, was equipped with a net drum to facilitate the handling of even the largest and most complex fishing gears. Unfortunately, unlike in *Cirolana*, none of this part of the vessel could be seen directly from the bridge, and the poor fishing mate had to rely on a CCTV screen to see the orchestra he was conducting.

If these had been *Scotia*'s only problems, she would have been far too busy to be used in exploratory fishing for lobsters. But the hydraulic system proved to be entirely inadequate to operate the main winches and the ship was laid up in Leith docks. After a time, alert Opposition Members of Parliament began to ask embarrassing questions of the hapless Secretary of State, making polite inquiries about how it was that the ship's company of a vessel permanently alongside continued to enjoy a distant-water allowance. Not long afterwards, I was offered the use of the ship and, with the enthusiastic support of her captain, Iain McBride of Dunoon, who had survived thirty-five days on a raft during the Battle of the Atlantic and was itching to get back to sea, we found ourselves steaming north, the decks piled high with lobster creels. Although the main winches were out of action, the anchor capstan was more than adequate to haul the largest fleets of creels and, with the help of the fishing mate, we devised a system whereby the long rope, to which the creels were attached by strops, could be led aft to a derrick so that the creels could be detached and handled on the working deck.

Not all of the ship's company shared my and the deck officers' enthusiasm. The chief engineer, Davie Pirie, a doughty child of the Costa Granite, was concerned at the numbers of scientists aboard (some of them were university visitors). 'The trouble, Dr Shelton, is that we're doon by the airse wi' fuckin' scientists' was one of his more printable comments. Like a number of the officers with Merchant Navy tickets, he preferred the more predictable life to be found aboard the civilian Fishery Cruisers (unarmed Fishery Protection Vessels) also operated by the Scottish Office. Scientific surveys, with their frequent changes of speed and direction, generated a lot of work for the engine-room staff, who rarely enjoyed the thrill of seeing unusual sea creatures coming aboard. So jaundiced was the chief's view of scientists that he often expressed the opinion that they were drawn from two classes as yet undescribed by sociologists, 'heid bangers and airse bandits'. But despite his regular protestations that key components of *Scotia*'s transmission system were 'a plate o' mince', and his insistence on a fifteen-minute shutdown of the main engines as we passed through the treacherous waters of the Pentland Firth, we were never held up by serious engineering failures.

On this, the first of a series of exploratory creel- and trap-fishing cruises with the crippled *Scotia*, we were able to confirm the presence of lobsters wherever the existence of rocky outcrops provided sufficient shelter from tidal streams. Our grid extended from the Butt of Lewis in the north to the islands of Barra and Mingulay in the south. The reefs around the St Kilda group marked our western limit. Our most important observation, from the fishery management point of view, was that we found no evidence of lobsters in the deep water to the west of the reefs. We had not yet explored the edge of the continental shelf or Rockall, but it looked as though there was no source of young lobsters outwith the area of the existing fishery. The fishery's future would depend, therefore, not upon larval manna from heaven produced by stocks as yet unfished, but upon the wise husbandry of the stock already known to the fishermen.

There was to be no easy solution to *Scotia*'s hydraulic problems. So long as the Scottish Office's steam side-trawler *Explorer* remained in service to undertake groundfish surveys, there was no immediate need to fix them. So it was that, a year later, *Scotia* found herself within sight of that remotest outpost of Inverness-shire, the Islet of Rockall. Not many people have heard of it and, of those that have, a surprising proportion think it's something to do with the weather. To call Rockall an islet is a bit like calling a Sinclair C5 a motor car. A steep cone of coarse-grained granite sixty-three feet high, it is no larger than many a Scottish tower house. Waves break over it often enough to sustain the growth of darkly pigmented diatoms on the surface of the granite. From a distance, this makes Rockall appear almost black, but on the day we were there the sea had been calm for long enough for sea birds to have streaked it white with their droppings.

Surrounded by a treacherous reef, Rockall is no place to take a thin-skinned steel research vessel. With *Scotia* barely under steerage way a mile and a half from the Islet, we struggle into our orange survival suits, mine at least a size too large and so stiff and new that it felt like rusted-up plate armour. Down the rope-ladder we swarm into the rubber Zodiac craft. The boatswain takes up position at the outboard, the bottom boards are piled with a dozen lobster creels, and First Officer Ian Macleod, my assistant Kenny Livingstone and I are crammed or perched wherever we can find a space.

The Rockall Islet, westernmost
outpost of Inverness-shire

It takes some time to reach the islet over the long, oily swell that passes for a flat calm in places like Rockall. By this stage, *Scotia* looks very small indeed. What the hell are we playing at leaving our secure home to wallow about among all these kittiwakes? What if the outboard gives up? I can't see any oars.

'Fuck' – we're alongside the towering islet, the sea sucking at its slippery sides as if licking a gigantic lollipop. Without hesitation, the gallant Kenny and Ian Macleod leap on to the rock. The Zodiac lurches, but the creels stay safely aboard as Kenny secures the rope, to which each creel is attached by a strop, to the islet. Ian and Kenny tumble back aboard and we 'shoot' the fleet, one creel at a time, over the side of the Z-boat, which now seems a lot roomier. 'Thank Christ. Let's get back to the ship ASAP.'

As we draw closer to *Scotia*, we see the rope-ladder rigged down the port side but, for some reason, Ian Macleod decides to go back aboard up the stern ramp. As he jumps out of the Z-boat, *Scotia's* stern briefly catches our bow and pitches poor Ian forward on to the steeply sloping steel ramp. Luckily, he succeeds in scrambling up it but, having seen the alternative, we elect to return by the way we left.

The open Atlantic can be an empty place, the straight-winged mollies (fulmars) scything over the waves our commonest companions with, every so often, a Rockall Jack (Arctic skua) or a bonxie, a kittiwake-bullying relative properly called a great skua, to keep us company. It's a different matter when trawling. Then, shrieking masses of kitties and other gulls appear apparently from nowhere to vie with the mollies and plunging gannets for the cod end's leavings. No such chaos enlivens our crab-fishing days. We spend the rest of the long hours of summer daylight hauling fleets of creels set from *Scotia* over the large sandy expanse of Rockall Bank some twenty or thirty fathoms below our keel. We catch numerous edible crabs, many bearing the black imprint of chitinoclastic bacteria, a sure sign of infrequent moulting and therefore of poor growth-rates.

Back on the bridge, we wait for the Mufax apparatus to spew out the latest weather chart. The stuttering pen sweeps back and forth, the clammy paper giving off a curiously anti-peristaltic smell, not a problem in calm weather but testing to the stomach muscles at other times. The high that has watched over us has hardly moved and, the following morning, the boatswain, Kenny and I are on our way back to the islet in the Z-boat. The creels are set among rocks but, as the last of the creels is handed aboard by the boatswain's mighty fists, our hopes of a Rockall lobster are dashed. Two dozen scabby-looking edible crabs scramble among our orange-clad feet, but nothing more exciting. We cannot be certain that there are no lobsters around the islet but it seems a fair interim conclusion, and we cannot afford to remain in the area any longer without risking the loss of the other fleets of traps (inkwell pots too big to carry in the Z-boat but capable of catching the largest crustaceans) we have deployed on Rockall Bank and down its eastern slope.

We found common edible crabs wherever we fished on the bank itself and down the slope to a maximum depth of almost 180 fathoms (still the deepest record for the species). Like the specimens from the islet, many were in poor condition by comparison with crabs taken on the North Sea grounds where feeding opportunities for bottom-living animals are much better. However, down the slope we encountered small numbers of edible crabs' gaudier deep-water relative, *Cancer bellianus*, known to fishermen as the lousy partan (Gaelic for crab) after Lousy Bank, the sea mount where they most often took it in their catches. All too soon it was time to leave for Aberdeen, yet it was not to be the end of the project. Using *Scotia* and, after the latter's eventual repair, the wonderfully seaworthy and reliable (even the winches were steam) *Explorer*, we fished for deep-water crustaceans on Rockall and Rosemary Banks, the Ymir and Wyville-Thomson Ridges and the edge of the Scottish Shelf to the west of the Hebrides. At around 300 fathoms we caught sufficient numbers of the delicious deep-water crab, *Chaceon affinis* (Milne Edwards & Bouvier), to convince ourselves that it might one day be a worthwhile fishery resource. We also persuaded ourselves that toughened glass had much to recommend it after *Explorer*'s over-refreshed galley boy had hurled a large rock at the front window of the Caledonian Hotel in Stornoway, only to see it bounce harmlessly back on to the street.

Over twenty years after my professional involvement in untangling the dynamics of lobster and crab populations drew to a close, it is gratifying to be able to report that these populations are still supporting sustainable fisheries, at a time when those for the once mighty cod family are not. Several special features protect them. The most important is that, in many instances, each lobster- and crab-fishing vessel sticks to its own patch, much as, say, a nesting pair of golden eagles has its own area of moorland which it defends against the competing claims of others. This territorial subdivision among the fishermen sets limits on the intensity of their fishing – limits which on exposed coasts like those of the Western Isles are reinforced by the impossibility of shooting and hauling creels and pots in heavy weather.

Just as there is competition among the fishermen, so the lobsters and crabs compete to enter the traps to get at the bait. As a result, the average size of the animals caught is greater than that of the population as a whole. The danger of catching too many lobsters and crabs before they have realized their growth potential is thereby reduced. Trap-fishing also confers another conservation benefit at the other end of the size-range. If the entrances to the pots and creels are too large, the animals that gain admittance can almost as easily escape. This constraint sets an upper limit to the size of the entrances. The very largest animals are too wide to get in and thus are able to contribute to reproduction year after year, safe from the effects of fishing.

Fishermen pursuing fish like cod, haddock and whiting in the three-dimensional immensity of the open sea are subject to no self-imposed territorial limit, and neither is trawl-fishing always efficient in releasing small fish. In this instance, only direct regulation and the imperatives of the market-place control the fishing effort. The result is that many of these fishes are caught long before they have achieved a fraction of their growth potential or indeed contributed to reproduction. When, as has happened to cod populations on both sides of the Atlantic, larval survival is reduced by climatic change, the stocks are no longer able to sustain themselves. Recovery from such a vicious perturbation of the marine ecosystem can take many years and poverty among both fishermen and fish processors is the inevitable outcome.

LAST IN THE FISH QUEUE

Such is the brevity of Man's written history, it is difficult to believe that, as recently as 12,000 years ago, most of the archipelago we now call the British Isles was not only connected to continental Europe but lay under many feet of ice. Only in the thin rump of southern England were there lakes and watercourses warm enough to support freshwater fishes. For instance, we know for certain that some bullheads, *Cottus gobio* L., in south-west England successfully survived the great ice age. Their DNA fingerprint is quite distinct from that of other British bullhead populations, most of which date only from the final retreat of the ice front some 10,000 years ago.

As the ice retreated, the fresh waters of Great Britain were gradually recolonized from the rivers of continental Europe and from the sea. The variety of my grandfather's fishing and the rich heritage of the River Chess owed much to the fact that the Thames and its feeder streams were once tributaries of the Rhine. The more the ice melted and retreated north, the more the melt water at both poles raised the level of the sea. At length, England's land-bridge to Europe was inundated and so there could be no more direct colonization by purely freshwater fishes like those of the carp family. Thus, although south-east England enjoys the most naturally diverse freshwater fish fauna in the British Isles, far greater ichthyological riches are to be found in the rivers and lakes of continental Europe.

Scotland was very much at the end of the fish queue 10,000 years ago. At first, only species with obvious marine affinities like lampreys, eels, whitefish, salmon, trout, charr and sticklebacks would have occupied the glacial rivers and ice lakes left behind as the glaciers fell back. It would have been a dramatic time. The retreat of an ice sheet is an awesome process which can be still viewed by passengers on the cruise ships visiting Glacier Bay in southern Alaska. During the long days of the northern summer, the glaciers calve great slabs of eerily blue ice into the sea below. Some of the falling slabs are taller than the ships' main masts and represent thousands of years of compacted snowfall. It is a gentler process inland, where the ice front usually has a softer edge and the melt water of summer trickles to form the glacial streams whose virgin waters will one day attract migratory charr and salmon straying north from established populations nearby.

We know all too little about the history of Scotland's fishes, but recent advances in the science of genetics have given us some insights into how salmon and trout populations re-established themselves. During the last and previous ice ages, both species occupied fresh and brackish waters far to the south and east which then provided similar temperature conditions to those in Scotland today. It seems that there was more than one of these so-called *refugia* in places like the Iberian Peninsula and North Africa. There was even thought to have been one in a glacial lake we now know as the Southern Bight of the North Sea. Genetic evidence has revealed how most Scottish salmon populations appear to be derived from the Iberian *refugium* and those of the Baltic countries and many of the Icelandic ones from the Southern Bight one.

At least two *refugia* appear to have contributed to Scottish trout populations but, unlike in salmon, some of this differentiation is recognizable in the appearance of the fish. The most widely distributed trout have relatively small numbers of largish brown spots, some of which also include vermilion. These fish are thought to be the most recent invaders. By contrast, in some isolated highland lochs, it is possible to find very different-looking trout with greenish-gold bellies and large numbers of very small brown spots. These fish are genetically distinct and appear to be remnants of an earlier invasion derived from a different *refugium*.

It could well be that other members of Scotland's indigenous fish fauna have similarly divided origins, and it is certainly true that, since their arrival, different populations of at least three species – trout, salmon and charr, *Salvelinus alpinus* (L.) – have diverged genetically to fit them better for the environments they currently occupy. The fact remains that, for all the complexity within species, Scotland's position at the top left-hand corner of Europe greatly restricted the number of different kinds of fish which arrived naturally. Even by the late eighteenth century, only another five species had been recorded. All had either adhesive eggs or a habit of forming egg strings at spawning time, so it is possible that they, too, could have been spread naturally on the legs and feet of waterfowl migrating to and from continental Europe. They were to be the last of the natural asylum seekers.

As time passed and first roads and canals and then railways made it easier for Man to transport fish and their eggs over larger distances, the fish fauna of Scotland became steadily but patchily more diverse so that, by the beginning of this century, the number of species breeding there was more than double that of the late 1700s. None of these upstart species – they were mainly members of the carp family brought up from England, grayling and a couple of North American salmonids – was to displace the Atlantic salmon as Scotland's premier game fish. So thoroughly has this relatively primitive bony fish entered public consciousness that it now enjoys cult status on both sides of the Atlantic. I was soon to remake its acquaintance but at a stage in its fortunes very different from the distant and bountiful days of the 1960s.

THE KING OF FISH

'Keep it up, Dick.' Two cock pheasants out of three had just fallen behind me in tribute to the genius of God's own gunmaker – Stephen Grant of 67A St James's – in turning an indifferent game shot into a passable one. Disabled by a stroke, his Purdey undisturbed in its oak and leather case, the laird still enjoyed his shooting as a spectator. Years before, he and his keeper had laid out the coverts that had made the estate into one of the finest pheasant shoots in Perthshire. I was there as his guest, having become his neighbour following my appointment as Head of the Freshwater Fisheries Laboratory at Pitlochry.

At first sight, not much at Pitlochry had changed since 1962, when I had first worked at the laboratory. The twenty-man barrack blocks, brought down in kit-form from Orkney in the late 1940s, still formed the basis for the accommodation in which chemists, biologists, office and library were all separated, to the great detriment of that day-to-day mixing of disciplines on which good science feeds. Despite this handicap, the laboratory had earned a worldwide reputation for its analytical expertise and for describing the ecological and physiological mechanisms that underlie the production of fish of the salmon family in fresh water. But if the facilities at Pitlochry looked much as I had remembered them, there had been important innovations elsewhere. With the enthusiastic support of the local salmon-fishing company, trapping and electronic counting facilities had been constructed on the River North Esk to make it possible, by an ingenious technique for tagging the young salmon, to follow the fate of successive generations of salmon throughout their lives, while a similar smaller facility had been created on the Girnock Burn on the Aberdeenshire Dee. Closer to Pitlochry, at Almondbank near Perth, water had been piped into tanks housed in a disused aeroplane hangar so that problems which seemed baffling in the field could be taken apart and studied in detail under controlled conditions.

The symbol of wisdom in Celtic mythology and the icon of conservationists on both sides of the Atlantic, *Salmo salar* L. still had

*An artistic but somewhat fanciful sketch of
the head of a salmon,* Salmo salar *L.*

the power to move the hearts of politicians and their officials. The
result was that the study of salmon had taken over the principal biological research facilities of a laboratory which had started life
studying brown trout. The new emphasis had not been entirely
welcome to the older staff, and even the most diehard of the salmon
enthusiasts regretted the replacement of the laboratory's wonderful
aquarium, with its live displays of the freshwater fishes indigenous
to Scotland, by a set of filing cabinets. The cold truth was that
Scotland's valuable salmon were in trouble and all the scientific
resources available had to be commandeered in their defence.

The salmon is incomparable as a food fish and the greatest trophy
to which a British angler can aspire, yet its wider appeal lies in the
heroic details of its life cycle. Like other so-called anadromous
fishes, salmon reproduce and undergo their early development in
fresh water but make most of their growth in the sea. By keeping a
foot in two quite different worlds, they ensure that during their vulnerable early life-stages they avoid the predators with which the
oceans swarm. Later in life, they take advantage of the much greater
food resources available in the sea to grow to a large adult size with
the reproductive potential to match. It is an ancient way of life by
no means restricted to salmon. In our own waters, sea and river
lampreys, *Petromyzon marinus* L. and *Lampetra fluviatilis* L. respec-

tively, sparling (smelt), *Osmerus eperlanus* (L.), and some populations of trout, *Salmo trutta* L., and sticklebacks, *Gasterosteus aculeatus* L., all take advantage of spending time in two quite different environments. What makes the Atlantic salmon heroic is that it takes this way of life to extremes.

Salmon, Salmo salar *L., leaping at the Falls of Muick on the Birkhall Estate*

The cycle begins when a hen salmon chooses a spawning site, usually at a ford where well-aerated water flows strongly both over and through the gravel. Some late-running salmon making for the lower end of big rivers like the Tay spawn within days of entering the river in January or February. For the earliest runners seeking a high burn at the top of the river, spawning may take place as early as October, the hen having spent as long as twelve or fourteen months quietly in the river. These earliest-running fish are the most venturesome. Many have migrated to and from the North-West Atlantic, only to be faced by another long upstream migration in the river. They break their upstream journey with resting periods in deep pools or lochs, and at these times lie virtually switched off in a state of suspended animation. Only in response to spates and in the final run to their spawning sites in late autumn do they expend appreciable quantities of energy in swimming against the current or ascending falls and rapids. Energy conservation is vital because, throughout their time in fresh water, the appetites of all sea-run salmon are completely suppressed by the process of sexual maturation, the earliest stages of which trigger their return from their subarctic feeding grounds.

Lay folk are often surprised at the coarseness of the fist-sized cobbles among which hen salmon construct their nests known as redds, but only by depositing their eggs among such stones can salmon be reasonably sure that they will not be washed away by the roaring spates of winter. Move the cobbles aside, as the hen salmon does when cutting a redd, and smaller pebbles and coarse sand will be found in their lee. It is in pockets of these lesser deposits that most of the eggs will spend the long winter months. But how can a fish which weighs no more than fifteen pounds, often much less, and has not fed for as long as a year, dig deep into the gravel to create a safe shelter for eggs in which thousands of miles of feeding migration have been invested? To a scientist digging into a dried-out redd to sample the eggs, or struggling with both hands to extract a frozen core sample to measure the depth of an egg pocket, the physical effort that lies behind the construction of the redd seems formidable indeed.

In fact, the hen salmon's achievement, impressive as it is, is based less on brute force and more on her exploitation of the properties of flowing water. When, in the half-light of dusk, she glides upstream from the deep security of the holding pool to the hurly-burly of the spawning ford, she precedes her cutting by turning her body to present the flat of her broad tail to the bed of the river.

Flexing her body vigorously, to create an area of low pressure below her tail and vortices alongside it, she creates a flow which pushes downstream the gravel and smaller deposits already buoyed up by the water they displace. Eventually, after periodic rests in slacker water, the hen fish creates a short trench roughly her own length and several inches deep, and checks its final depth with the sensitive tip of her anal fin.

Throughout this activity, minute quantities of pheromones present in the ovarian fluid occasionally leak from her vent and attract cock salmon from below. Sometimes competing aggressively to win her and sometimes in uncontested possession of his dusky love, the dazzling male stands by. The carotenoid pigments accumulated from his marine feeding have been liberated from the fat he has used up during his own months of starvation and redeposited to give his skin a speckled orange brightness that doubtless intimidates his competitors. No such gaudy bad taste mars the understated beauty of the hen, most of whose carotenoid pigments are now locked safely away in the yolks of her eggs. Although much darker than when she first entered the river, she never entirely loses her silvery sheen, which is sometimes enlivened by the merest hint of violet.

A bright livery is not the only aggressive aspect of the cock salmon's appearance. As his sexual development has proceeded, his skin has thickened, the adipose dorsal fin in front of his tail has increased in size and his upper and lower jaws have enlarged grotesquely to form the permanent snarl of his kype. All in all, his fifteen pounds of menace should ensure that he is given a wide berth by smaller sea-run cocks. But they are not to be his most serious rivals. The pheromones that have attracted him have also

stirred the blood of a group of male salmon parr, rampant minia-
tures five or six inches in length which have matured in fresh water.
So richly have they been feeding in the river during the previous
spring that the increased rate of change of day-length around the
equinox has triggered sexual maturation without the need for sea
feeding. Now they sport about the vent of their giant madonna,
oblivious to the snapping kype of their mighty sea-run rival whose
jaws will kill or grievously wound a few of the less fortunate
among them.

The quivering displays of the big male stimulate the exquisitely
sensitive organs of touch at a distance within the lateral-line organs
of both himself and his mate. Soon both are open-mouthed in a
joint orgasm, and eggs and milt (sperm) are discharged simultaneously into the bottom of the redd. It is an orgasm in which the
eager parr also join. So successful are their endeavours to pulse their
milt into the very depths of the redd that, when we come to sample
the DNA in the fertilized eggs, we will find that nearly half are the
children of parr. The eggs have been shed into several pockets in
the hen's one successful redd – she has abandoned two others
nearby when she encountered bedrock – and her final spawning
act has been to cover the eggs by cutting into the gravel deposits
upstream so that they cascade gently back to form a stony blanket.
A few eggs have been swept away and trout and salmon parr have
eaten others, but most of the 5,000 are now safe below the turned-
over gravel, showing bright and clean against the undisturbed
sediments to either side.

Her duty done, the hen kelt or spawned-out salmon drops back
downstream. As she shelters in the main stem of the river, her weak-
ened body attempts to repair itself. She is still alive in the spring and

her dusky sides have taken on the bright chromium plate of a mended kelt. Slipping back to sea, she sees the seal briefly, but the burst speed of her starved body is no match for that of her assailant and, like most hens and almost all sea-run cock salmon, her first romance will be her only one. As for her paramour, he remains in the spawning area, dashing about in further attempts at mating and expending his remaining energy in futile battle with other cocks. By the end of it all, his immune system, already suppressed by the corticosteroid hormones driving his aggression, has lost the fight with the bacteria and fungi that infest his skin and he dies gasping in the shallows as a dog otter bites deep into his watery flesh.

The hen has chosen her redd site well. Throughout the winter, well-oxygenated stream water bathes the egg pockets and, by the time the spates of February come to rattle the gravel, the embryos developing within the eggs are robust enough to withstand the disturbance. Hatching as alevins in the early spring into the gravel of the redd, for a time they still enjoy the protection which their mother's instinctive spawning behaviour has conferred. It is to be the last of her gifts; the world into which the swim-up fry emerge in April and May provides new opportunities but also terrible dangers. For those few that secure feeding territories before the maternal legacy of sustenance in their yolk sacs runs out, their emergence marks the start of a veritable odyssey. For most, however, life ends with independent existence. As for the initial survivors, the great English fishery scientist David Cushing put it, 'Fish grow to avoid mortality.' The longer a fish is small, the more likely it is to die and, by the end of the first year, less than 10 per cent of the young fish which leave the redd have found sufficient food and survived predation by mammals, birds and other fish.

As the growing season progresses, so the food requirements of the young fish increase and the larger of them seek feeding territories in deeper water farther out from the river bank. With the greater flow comes a greater supply of drift fauna. Some, such as mayfly and stonefly larvae, are indigenous to the river, while other organisms are blown in off the sedges, meadowsweet and alder bushes along the bank. By this stage, the young fish have acquired the thumbprint-like parr markings along their sides which help to break up their outline against the stony bed of the river. Although they take advantage of strong flows, salmon parr do not waste energy in swimming against them. They spend most of their time in the boundary layer of slower-flowing water near the bed or in the lee of large stones. In both instances, they use their unusually large pectoral fins as hydrofoils to help them remain in position with a minimum of active swimming. This they save for such special occasions as seizing prey from the flow above, chasing rivals and escaping predators.

The lives of parr can follow a number of pathways. The fate of almost all surviving females is one day to leave their feeding territory to join with small shoals of other parr of both sexes on their journey to the sea. In Scotland, reaching this stage in the life cycle can take anything from one to four years, but most salmon do it in two or three. The earliest downstream migrants of both sexes leave in autumn from the upper parts of the larger rivers and they join young fish from lower down the system to enter the sea in spring. Male parr also form important components of the downstream movements in autumn and spring but some remain behind to spawn. In so doing, especially if they do so over more than one breeding season, their chances of joining the migration to sea are reduced. Preparing to breed and preparing to enter the sea present conflicting physiological demands, so much so that those of the former tend to suppress the latter. This conflict reduces the numbers of previously mature parr that are able to migrate to sea in the spring following their participation in breeding. Breeding is also dangerous. The raised levels of sex hormones reduce the resistance of the fish to disease. Furthermore, leaving shelter to hang around the vents of female salmon renders the parr more vulnerable to predation of all kinds and to attacks from the large sea-run cock salmon.

The fish that enter the sea each spring are known as smolts, a word whose sound and very derivation reflects their glittering appearance of molten metal. Their silvery coat of guanine crystals, which mirror the sea around them and thereby provide a form of camouflage, conceal the parr marks beneath. Although it is the most obvious adaptation to convert a fish which hugs the bed of a river into a member of a shoal which swims near the surface of the sea, the acquisition of a metallic overcoat is only one of a number of complex changes that together form the process of smolting. Fins and tail darken almost to black and the fish becomes elongated to fit it better for continuous swimming. Most important of all, in exchanging the medium of fresh water, which is much more dilute than its own blood, for that of the sea, which is more concentrated, the young salmon encounters all of the problems of a desert mammal. In fresh water, it avoids the potentially fatal problem of oedema by excreting large quantities of water via the kidneys. In the marine environment, however, it can avoid lethal loss of fluids only by drinking sea water and excreting the salts via the gills and gut.

Complex as the behavioural, physiological and ecological mechanisms are that end each spring in pouring young salmon into the sea, the results of our work at the Girnock Burn, and those of other workers across the Atlantic salmon's range, showed that the summary mathematics were remarkably simple. Depending upon the abundance and distribution of spawning salmon, the numbers of smolts which leave each year are limited either by the numbers of eggs laid or the carrying capacity, in terms of food and shelter, of the river. In a sense, therefore, safeguarding a salmon stock from the total effects of all forms of fishing is a relatively straightforward matter: so far as is possible, enough fish must be left to spawn to ensure that subsequent smolt production is limited by carrying capacity rather than egg supply.

Interestingly for me, with my recent experience in trying to unravel the population dynamics of lobsters, there were direct parallels between the lives of these large crustaceans and those of salmon. Provided the hen lobsters release enough larvae into the sea from the eggs glued, like so many berries, to the pleopods under their tails, the numbers of young lobsters entering the fishery five to seven years later depend upon the capacity of the seabed inshore to provide shelter and nurture for the settled larvae. For both species, the final arbiter of juvenile abundance is the territorial extent of an area, which can support only so many young animals. Add more, as happens when spawning stocks are high, and the result for both salmon and lobsters is merely greater mortality for the early life-stages as they compete for places in a limited environment. It is not the only parallel between the life cycles of these two totally unrelated species. Just as the lobsters which join the fishery each season comprise the products of several successive years of spawnings, so does the population of salmon smolts which enters the sea each spring.

For a number of reasons – some complex and best left for the consideration of fishery scientists – variations in the strength of salmon generations tend to be much less than for other species of fish. None the less, as so often in biological science, a little knowledge can be a dangerous thing. We soon learned that a river's carrying capacity varies both within and between years as a result of changes in factors like rainfall, temperature and land use. Such variation could be allowed for. Less easy to accommodate was the increasing realization that homing species like Atlantic salmon tend to split into separate populations with important differences in run-timing and in the triggering of smolting and sexual maturation. And given that most fishing took place far from the parts of rivers to which the adult salmon home, it became obvious that it would be difficult to control the levels of that fishing to ensure that adequate numbers of spawners remained to sustain all of the separate populations.

There was a further complication. The migratory behaviour of salmon renders them vulnerable to a hierarchy of exploitation: there is a succession of fisheries which depletes the stocks at various points in their feeding and spawning migrations. Unsurprisingly, great and long-standing jealousies exist between the participants in these various fisheries. Muted at times of abundance, these jealousies become more strident at times of shortage. They became most strident of all with the growth, during the 1960s and 1970s, of high-seas fishing for salmon at West Greenland and Faeroe by drift net and long line respectively, and of large offshore drift-net fisheries in the coastal waters of Ireland and north-east England.

To a visitor from another planet, the fact that so much scientific and political effort should be devoted to the management of a single, rather uncommon fish might seem inexplicable. Even during periods of relative abundance, the contribution which wild Atlantic salmon made to the national food supplies was minuscule. Yet now, at a time of scarcity for wild salmon, but of unprecedented supplies of farmed ones, interest and investment in the wild stocks has never been greater. Part of the explanation is the high value that anglers place on the opportunity of catching a salmon, but it is not the whole story. Environmentalists esteem this fish just as highly, as a symbol of the ecological health of rivers and of our stewardship of the sea's bounty – a bounty in which salmon also share. For a developed nation to fail to look after its wild salmon is now seen as almost as great a badge of shame as to contribute to global warming or to the acidification of rain. It is for this reason that countries devote whole laboratories to the monitoring of wild salmon resources, and fishery scientists spend entire careers researching their biology.

SALMONID ORIGINS

For many years, naturalists have divined the probable relationships between animals by comparing the structures of living creatures with one another and with those that have left their fossil remains in the rocks. It is a patchy and uncertain process, dependent in turn upon the range of relatives the ruthless forces of natural selection have bequeathed us and on the chances of finding relevant fossils.

We know from such features as the arrangement of their fins and the fact that their swim bladders open into the pharynx that salmon are members of the old brigade of modern bony fishes, a group that became established nearly 200 million years ago. Nevertheless, despite the evidence of antiquity spoken by their anatomy, the most recent fossil recognizable as a salmon is only some five million years old.

The explanation is not that salmon are recent products of evolution, merely that their life-style as juveniles in torrential rivers and as maturing adults in the surface waters of the open ocean is inimical to fossilization. With the advent of the new science of cytogenetics, the study of the ways in which genetic material is packaged in cells, came the realization that the cells of the most primitive members of the salmon family had remarkably large numbers of chromosomes (ninety-six to our own forty-six). It was clear that, at some point in the history of the group, there had been an accident during reproductive cell division that had doubled the number of chromosomes in the next generation. Such polyploidy is quite common in flowering plants, where it is often associated with rapid growth to a larger size. How would such a chance event have been helpful to the proto-salmon?

Enter the smelt or sparling, *Osmerus eperlanus* (L.), a small silvery estuarine fish closely resembling a salmon in miniature and which spawns above the head of tide (like salmonids, it can complete its life cycle in fresh water if land-locked) but makes most of its growth in brackish and coastal waters where food is more abundant. In structure and life cycle here is a fish which appears to give important clues to the kind of creature from which the majestic salmon evolved and – guess what? – it has only about half of the proto-salmon's genetic material. We have no way of knowing for certain whether the first of the latter was a genetically doubled-up version of a fish like a smelt. However, for a fish which extended a brackish water life-style to the even more productive mixing zones of sub-polar seas, a spot of polyploidy might well have been strongly advantageous in enabling rapid growth to a large size and, consequently, the reproductive power to out-compete the less well-endowed coastal opposition.

In a nutshell, it looks as though the life cycle of the salmon of today is that of the smelt's extended upstream in fresh water to reduce predation on eggs and young and far offshore to take advantage of the lavish feeding opportunities in the subarctic. How long did it all take? We do not know, but geneticists speculate that the transition could well have taken place some sixty million years ago. That would have been about halfway through the history of the Atlantic Ocean itself. Just how the complex navigational abilities evolved that enable salmon to undertake their long sea migrations and return to their natal rivers we will probably never know. Certainly, the task was less awesome when the Atlantic was young and relatively narrow but to me, at least, it is still a source of wonder.

KILLING THE KING

For most of Scotland's recorded history, her salmon fisheries were confined to the rivers and estuaries to which most of the fish were homing. Ascending fish were taken in sweep nets, so-called fixed engines of wood and stone (cruives), leistered (speared) by the light of torches of burning heather, and angled by rod and line. As with most artisanal fishing – especially artisanal fishing under the cold eye of feudal superiors – these primitive methods were relatively inefficient and so rarely was the stock as a whole, or the individual populations of salmon which comprised it, put in mortal danger. Although the quality of the fresh-run fish was good enough, provided they had not been too roughly handled, the quality of salmon which had been in the river for some time – and had consequently used up part of their fat reserves in day-to-day existence and the development of eggs or milt – left much to be desired. This loss of quality was especially felt in the catches of the leister men, who often achieved their best results with salmon on the point of spawning.

As the Industrial Revolution progressed and the new railways gave access to the developing markets for fish in the cities, so better methods were developed for catching salmon in good condition. The eating qualities of the fish were known to be at their best before they had even entered the rivers. The concept of the 'fixed engine' was therefore transferred to the open coast, but the technology was to be very different. In seeking the scent of their home river, salmon and sea trout swim for considerable distances close inshore. By siting a wall of netting, or leader, at right angles to the shore, it was found that the fish could be diverted into a succession of so-called netting inscales and thence into a fish court from which the catch could be extracted with a long handled hand net.

The coastal fixed engines took various forms to suit particular sites. On rocky coasts, it was usually necessary to moor the entire structure, including the leader. These bag nets were fished by brave men working from broad-beamed cobles with upswept bows and a square transom. On beaches, it was possible to stake the net directly into the sand and these fly nets (in which the entire net, including the leader, is staked) and jumper nets (in which the leader floats into position with the incoming tide) became an established feature of the east coast of Scotland. Their iced catches were to set a standard for the culinary excellence of Scottish salmon – one on which the rather different product of the cultivation industry continues to trade.

Salmon netting in the nineteenth century.
The circular hand net is known as a skum.

To salmon biologists, the catches of the coastal fixed engines were to prove a useful source of fish for tagging. Using simple external tags, it was possible to estimate the levels of exploitation by the fixed engines themselves and by the subsequent net-and-coble and rod-and-line fisheries in the rivers. With the invention of the acoustic tag, it was even possible to follow the routes of individual fish migrating along the coast and, often successfully, avoiding the fixed engines set to intercept them.

Netted salmon are normally sent to market packed in ice, their snow-white bellies uppermost and unsullied by the gutting knife. In preparing such fish for the table, the guts were usually found to be empty but sometimes, in the early part of the fishing season when testes and ovaries are at their least developed, the remains of sea feeding, especially upon sand eels, *Ammodytes marinus* (Raitt), could be recognized in the stomachs. It was a reminder of how little was known about the lives of salmon at sea. The advent of large-scale high-seas fisheries for salmon in the 1960s and 1970s was shortly both to extend that knowledge and to underline the importance of obtaining it.

For all of the opportunities coastal fixed netting stations offered to the cause of science, their advent as a fishing method represented a backward step for the safe, sustainable management of salmon populations homing, not just to individual rivers, but even to particular parts of the larger ones. Netting fish on the open coast intercepted them before the populations had fully separated, and so matching spawning stock to a river's carrying capacity became correspondingly more difficult. In times of abundance, such concerns are more theoretical than real, but when stocks are low, the dangers of depleting the weaker populations by this form of fishing cannot be ignored. But if there were worries about the coastal fisheries, with their comparatively local effects, there was much greater concern when new fisheries grew up far from Scotland. The fishery at West Greenland exploited the stocks of salmon returning to both Europe and North America, and another, around Faeroe and in the Norwegian Sea, intercepted salmon from European rivers.

Kapisillit is the Greenlandic word for salmon, and also the name of the one river in Greenland where, in most years, salmon spawn. That salmon also occurred in fishable numbers off the coast had been known to the Inuit for many years, but it was not until the 1960s that serious worries began to surface about the home-water origins of the catch. By this time, the growing native gill-net fishery had been supplemented by a large-scale offshore drift-net fishery using the latest monofilament nylon material. The total landings first exceeded 2,000 tonnes in 1969 and did not fall below this figure until 1976 when, under the terms of a quota agreed in 1975, following bilateral discussions between Denmark and the United States, catches were limited to 1,190 tonnes. Sampling of catches showed that, in most years, the home-water origins of the West Greenland salmon were about evenly split between the rivers of North America and Europe. Furthermore, the catch consisted almost entirely of fish which, had they survived, would have come back not as grilse (one-sea winter salmon) but as substantial multi-sea-winter salmon, many of them large females returning early in the fishing season and the most important source of eggs for the upper parts of river systems. Over time, the stocks off Greenland have fallen drastically. The fishery has now been reduced to what is effectively a subsistence-level activity.

During the 1970s, an equally undesirable high-seas salmon fishery, this time prosecuted by baited long line, developed around the Faeroe Islands and in the Norwegian Sea. At its peak in 1981, over 1,100 tonnes of salmon were taken in this fishery, all of them potential recruits to European rivers. As the fishery moved north of the Faeroe Islands, so an increasing proportion of the catch consisted of potential multi-sea-winter salmon. As with the Greenlandic fishery, that at Faeroe was later limited by annual quota. After a few years during which the quota was bought out by angling interests, the fishery is now controlled directly by the Faeroese Government. In recent years only a small research fishery has been permitted.

Sadly, there was to be no immediate relief from the depredations of the large drift-net fisheries which also developed off the west coast of Ireland and the north-east coast of England. Irrespective of the terms of the treaty which set up the North Atlantic Salmon Conservation Organisation, these fisheries continue to exploit the villainous properties of the mono- and multi-monofilament drift net, killing fish, birds and marine mammals without discrimination. However, there is light at the end of the tunnel. Fifty-two of the sixty-eight Northumbrian and Yorkshire netsmen have agreed to exchange their licences in return for gratuities to which both private and public funds have contributed. Controlled reduction of the Irish drift-net fisheries is also in prospect, perhaps with the eventual possibility of a buyout.

As it was, the advent of large-scale drift-net and long-line fisheries coincided with a period of unusually high abundance for wild Atlantic salmon which extended from the early 1960s into the early 1970s. Fish of the cod family (Gadidae) also enjoyed exceptional abundance over the same period – the so-called gadoid outburst. Salmon, cod and haddock are all cold-water fishes which ultimately depend on the marine feeding opportunities generated by the mixing of the Arctic and Atlantic oceans. Moderate cooling of the Northern Hemisphere enhances this process and the fish benefit.

But the 'mini ice age', as the climatologists called this exceptional period, was not to last and, by the late 1970s, its beneficial effects on the early marine survival of salmon, and of other subarctic fishes, had come to an end. Of the British salmon populations, the worst affected were the early-running ones, and these were already bearing the brunt of a devastating outburst of that enigmatic disease of maturing salmon, ulcerative dermal necrosis

(UDN). Skin infections are common in maturing Atlantic salmon in fresh water. This is mainly because the high levels of steroid hormones that drive sexual maturation also tend to suppress the fish's immune system. Usually, however, such infections are not fatal until after spawning. UDN was a much more virulent condition in which sores on the head spread rapidly over much of the body of the fish, acquiring as they did so a veritable blanket of bacteria and fungi. The causative organism was never identified, but most pathologists now believe it to have been a virus. It took some years for the outbreak to die out, just as it had in a similar instance some sixty years before. In the meantime, resources of early-running salmon were being catastrophically depleted at the same time as they were suffering losses from high-seas fishing.

Many of the older members of early-running salmon populations make for the North-West Atlantic in the later stages of their feeding migration. The reduced numbers of these fish were mirrored in those of the North American (largely Canadian) salmon, whose feeding grounds they shared. Similar problems started to affect the survival of salmon feeding in the North-East Atlantic, and catch-rates started to fall in both the high-seas and home-water fisheries. The good times were emphatically over, and environmentalists in Europe and America began to express serious concerns about the future well-being of the fish they had made their icon.

In wild fisheries for most species, reduced catch-rates are normally accompanied by increases in unit price. These increases enable fishing to continue at a profit, even when the underlying abundance of the resource is endangered. This is always an undesirable process, yet in the case of wild salmon it was to have minimal effects. Techniques had been developed for farming salmon intensively and cheaply enough to depress its price as a commodity, and as a result the total abundance of *Salmo salar* was to be raised to levels unprecedented in the long history of the species. No longer was the world supply of Atlantic salmon to depend on the vagaries of the wild stocks.

CHILDREN OF
THE SPINDRIFT

It had been Jamie's father, the Captain, who had introduced him to Loch Maree. A veteran of the Battles of the Dogger Bank and Jutland, the Captain had left the navy in the late 1920s to run the family estates in Hampshire and he had first seen the loch at the invitation of a fellow member of HMS *Birmingham*'s wardroom. Now, as he sat there in the boat with his newly commissioned son on a mild June day in 1939, it hardly seemed possible that once again the grey ships would steam out from Scapa to confront the same enemy who had stolen his youth just two decades before.

There was a good ripple on the loch. Foam, dead grass stems and the occasional struggling insect marked the long windrows and, as the taciturn ghillie Angus John rowed the heavy boat upwind to begin the first drift, both father and son fancied they saw the blue nebs of sea trout breaking the troubled surface. Out with the drogue to slow the boat's progress before the wind and the drift began. The ribbon-like floss lines streamed from the tops of the long greenheart dapping rods and, at the tip of each gut-cast,

The light cruiser, HMS Birmingham *(Captain A. A. M. Duff RN), veteran of the Dogger Bank and Jutland.*

a bushy fly sported among the windrows. Now on the water, now swept above the miniature white horses by the gusts, the flies were soon noticed by the sea trout. The Captain's was first to disappear in a fierce boil that jerked the cast as taut as a bowstring. Had the rod tip not been well raised to cushion the rush, neither fish nor fly would have been seen again that season. Within a minute, the floss was down to the backing.

Yelling to young Jamie to reel in his line, Angus John scrambled the drogue aboard and pulled the boat round with arms strengthened by many a season at the peats. A dozen powerful oar strokes in the direction of the fish boring deep below, and the tackle survived the first crisis. No sense of tugging was transmitted to the Captain's gnarled hands, only a continuous pull that neither he nor Angus John could resist. Both knew that they were into a good fish, and the Captain was secretly glad that it was the veteran ghillie and not himself who had tied on the fly and sharpened its hook with his whetstone. Suddenly the pulling stopped, the line slackened and a quivering silver missile shot vertically out of the water. With equal speed and without conscious thought, Angus John reached across to push the Captain's rod down and sideways so that the raging leviathan would not fall on a taut cast and snap it at the ᴸ indeed, it is a big fish, Captain,' said the ghillie with the beautifu. modulated formality of one whose first language is Gaelic.

It was to take half an hour before the fresh-run sea trout was alongside the boat and experienced hands drew it inboard in a single decisive stroke of the long-handled net. The ghillie reached for the nabby (a leaded truncheon) lying on the foot boards. 'No, Angus John,' said the Captain, 'a big fish like this will lay a lot of eggs. Let's slip her back and hope she lives for another year.' Smiling quietly, Angus John wetted his hands and lifted the exhausted madonna over the gunwale. At first she lay still, her gills working hard, and seemingly unable to stay upright without support. The build-up of lactic acid in her muscles would take many hours to dissipate, but within a few minutes she had recovered enough for her great square tail to take her deep into the dark waters of the loch.

At last the sub-lieutenant found his voice. 'Why on earth did you put it back, Dad? It must have been a twelve-pounder at least.' 'I know, Jamie,' came the reply, 'but she must have spawned four or five times at least. The flesh of fish like that can be pale and chewy. Better that we remember her here in the loch than as a sad disappointment on the dinner table.'

It was to be a day of days on Loch Maree. Both men caught fish and the taste of the simply cooked three-pounder they finally took back to the hotel could not have been bettered. Jamie left to join his ship the following morning and he was not to see Loch Maree again for seven long years. The Captain, recalled to serve aboard vulnerable merchantmen as a convoy commodore, was not to survive the war. His ship was sunk by a U-boat and, on a grey afternoon in the third winter of the war, he lost his grip on the Carley float. As he slipped below the oil-streaked sea, his son could only stare into the spindrift from the open bridge of his corvette as it plunged ahead on the starboard wing of the convoy.

The sub-lieutenant of 1939 was now a retired commander obliged by his family responsibilities to leave the service he loved and follow his father's footsteps ashore. Angus John's time in the Lovat Scouts had come to an end not long before and, when the two men met, equals now, at the boatshed, only unspoken thoughts of the old Captain dimmed the joy of their reunion. 'The fish are waiting for you, Jamie' was Angus John's greeting. It was to be the first of many such meetings. When there was wind and water, they did very well, and when there was not, they spent less time on the big loch and more on the high hill lochans among the bright brownies. Always there were fish to be had and every so often a veritable submarine would bring to mind the Captain's fish.

Jamie was in the study of the big house when the telephone rang in 1989. It was Angus John. 'The fish are just no' here, Jamie' was his anxious message. Right enough, they had not seen as many over the last year or two and they had not seen a double-figure sea trout for several seasons. But similar hiccups had happened before at times of low summer rainfall and neither of them had worried unduly. Despite Angus John's entreaties, Jamie had no intention of cancelling his annual fishing holiday but, during the long run north in the train, he began to think about what might lie behind Angus John's message. Jamie was not a great reader, but he kept up with the sporting press and had heard about the problems sea trout had been experiencing in County Mayo and Galway. Not only had their numbers fallen sharply, but there were reports of fish returning early to fresh water grievously scarred about the head and dorsal fin. The Irish scientists who had sampled the stricken fish in the Burrishoole river they had been monitoring for many decades, were certain that the injuries were caused by the same sea lice that were infesting the caged salmon in farms along the coast.

It was a shaken Angus John who collected Jamie at Inverness in the battered old shooting brake. The day before, he and a couple of fellow ghillies had been speaking to Dr Andy Walker of the Freshwater Fisheries Laboratory at Pitlochry. Dr Walker's careful sampling had convinced him that sea trout populations up and down the west coast of Scotland were in mortal danger from the very same cause he had now seen at first hand when visiting his fellow scientists in Ireland.

Jamie caught a few finnock (trout that have returned to the river after a few weeks at sea) during his fortnight and one sea trout of about a couple of pounds. There was no summer job for Angus John the following year and, for the first time since the war, Jamie's dapping rod lay untouched in the gunroom. As season followed season, and the caged fish grew ever more numerous, the more obvious it became that the new jobs on the salmon farms had been bought at a terrible price. The sea trout had been the first to suffer because they spend the summer feeding close inshore, the very place where sea louse larvae tend to accumulate. Soon there were complaints from rivers entering long sea lochs that the salmon populations were also in trouble as their smolts ran the sea-louse gauntlet. Unlike the sea trout, salmon smolts are rigidly pro-grammed to stay at sea until they reach adulthood. Theirs was a hidden Calvary, far offshore, as the maturing lice they had picked up in the sea lochs bit through the skin and destroyed the fluid balance of the worst-affected fish. It was to be years before a

reluctant government would grudgingly accept the disastrous consequences which their regional development policies had helped to
create. With blooms of algae in mid-water that forced the seasonal
closure of valuable scallop fisheries, with vile blankets of rotting
ordure below the salmon cages, the sheltered waters of the west
were an undersea Eden no longer.

THE BLACK HOLE
IN THE SEA

'Where the hell are all the salmon, Dick?' As the twentieth century
entered its last decade, so the question became more insistent.
Granted that there were worries on the north-west coast of
Scotland, where in some rivers it seemed that there were no longer
enough spawners to stock them with young. But all the indications
were that the big east-coast rivers were sending as many young
salmon to sea as always. The problem lay with the proportion that
returned.

Perhaps the stocks had been contaminated by interbreeding with
escaped farmed salmon poorly adapted for independent life in the
ocean. There was some evidence that such mongrelization was
taking place in north-west Scotland and even that some escaped
female salmon wasted their eggs by allowing themselves to be
seduced by male trout. Every so often, escaped farmed salmon
would turn up in rivers like the Spey, Dee, North Esk and Tay, but
in numbers so small that serious interbreeding problems seemed
highly unlikely as an explanation for the lower survival of their

smolts in the sea. We also found no evidence for the idea that milder winters had in some way reduced the ability of the smolts to cope at sea by interfering with their capacity to face the transition from fresh to salt water or to mistime their migration down-river. No, the problem seemed to lie in the sea itself. Once again, it was time to ask the fish.

Because the numbers of wild Atlantic salmon are restricted by the amount of food and space available for their early life-stages in fresh water, they have always been rather rare fish in the sea. They are at their most numerous in the first weeks after leaving the relative security of the river and before the main effects of sea mortality have taken their toll. Our earliest efforts to catch the young salmon at this time had been with small-mesh gill nets set off the coast on what we imagined were their migration routes. We caught a few sea trout, sprats, *Sprattus sprattus* L., and three-spined sticklebacks, but never a post-smolt salmon. The reason for our failure emerged at an informal workshop on salmon in the sea organized by the Atlantic Salmon Trust, a charity devoted to the well-being of wild Atlantic salmon and sea trout.

One of the participants in the workshop was the Norwegian biologist Dr Marianne Holm and she brought with her a videotape of hatchery-reared salmon smolts which had been released into a fjord. The fish swam together near the surface in a loose shoal. This was interesting in itself, but even more revealing was the smolts' reaction to a gill net in their path. Instead of blundering into it, they had no apparent difficulty in detecting and swimming around it. We abandoned any further thoughts of trying to catch post-smolt salmon with gill nets and, the following year, observed the behaviour of hatchery-reared and wild smolts in a netted-off enclosure within a large tank of sea water pumped directly ashore from Loch Ewe in Wester Ross. Both groups of fish swam mainly at the surface, but though the reared smolts formed a loose shoal, like Marianne's, the wild fish, which had been trapped at the head of tide in the River North Esk in Angus, clung together in a very tight shoal indeed. Two of the smallest wild smolts inadvertently slipped through the mesh of the netting enclosure yet, so strong was their instinct to aggregate that, instead of swimming away to apparent freedom in the wider world of the tank, they made frantic efforts to return through the mesh to join the rest of their shoal.

Like ourselves, the Norwegians had received a number of reports of young salmon being caught accidentally in mid-water trawls. In our case, the reports had come from herring fishermen working in the Firth of Clyde who provided scattered records of post-smolt salmon, some of which we had reared and tagged ourselves at a facility we were operating on the Mull of Kintyre. Our Norwegian colleagues also received similar reports from their herring fishermen, but the real breakthrough, if such it can be called, came when mackerel fishermen discovered that they could often increase their catch-rates by modifying their mid-water trawls to fish at the surface. By-catches of salmon became much more common. Occasionally, returning adults were taken, but most of the salmon were post-smolts similar in size to mackerel and easily confused with them once the scales had been rubbed off by contact with other fish in the cod end of the net. Often considerable numbers of post-smolts were caught in this way, and it was not long before the Norwegian scientists were able to confirm, by their own exploratory fishing, what we had seen in our tank experiments: namely, that post-smolt salmon tend to form small shoals that spend long periods of time close to the surface of the sea. It was in the light of this knowledge, and of the positions where the Norwegian scientists had caught salmon, that we were able to plan some exploratory fishing of our own.

''Ow do, ye fookin' basstud' – the boatswain's friendly greeting, delivered in the unmistakable tones of Hull's Hessle Road, marked a great change in the life of the old ship. *Scotia*, aboard which I had ventured to Rockall so many years before, was now operated by Messrs Marrs, the famous Humberside trawling and shipping company which had won the contract for managing the Marine Laboratory's remaining two research vessels. The ship itself looked rust-streaked but otherwise much as I remembered it. Up forrard, the slab-fronted bridge structure loomed as large as ever, as helpful to forward progress against a gale as a man-o'-war's mains'l taken aback. On the working deck aft, all was as before, at least in appearance. The hidden difference now was that the hydraulic system had been completely replaced and, for the first time in the life of the ship, everything now worked. With trawl doors clunked into position on the stern blocks and the big pelagic trawl safely furled on to the net drum, we were 'ready in all respects for sea'.

After working two lines of hydrographic stations in the North Sea, we were on our way to the Faeroe-Shetland Channel in an attempt to capture post-smolt salmon migrating north-eastwards in the surface waters of the Shelf Edge Current to their summer feeding grounds in the Norwegian Sea. Our first trial with the big pelagic net ended in disaster when it fouled a rocky outcrop to the east of the Orkneys, but the skills of the Marrs' men soon had all put to rights and the second trial farther offshore worked perfectly.

Full of hope, we steamed north to the Shetlands but, just as we were looking forward to trying our specially rigged trawl in anger, such a gale blew up that we were obliged to shelter the storm-battered *Scotia* in the lee of Unst, the northernmost of the Shetland group.

Thirty-six hours later we were on our way and, by the following dawn, the huge net was streaming astern until, with the warps – the multi-stranded wires that connect the trawl doors to the ship – fully extended, we were fishing in earnest. Two hours' towing at three knots – the relatively high trawling speed necessary to catch young salmon – yielded a few mackerel, *Scomber scombrus* L., five bizarre-looking bathypelagic refugees from the depths and a large number of young lumpsuckers, *Cyclopterus lumpus* L. Plain chocolate brown above and a delicate pale turquoise below, the 'paddlecocks' were to appear in nearly every haul and, unlike the mackerel, a huge porbeagle shark, *Lamna nasus* (Bonnaterre), the herring, *Clupea harengus* L., and the green-boned garfish, *Belone belone* (L.), which were our usual by-catch, the lumpsuckers survived their cod-end adventures well enough to swim away with no apparent ill effects.

My self-appointed sea-duty station during hauling was on the starboard quarter, tucked between the groaning and crackling drum of one of the two trawl winches and the gunwale. Here, out of everyone's way and in mute defiance of the great gods Health and Safety, I had a grandstand view of the returning net as it hissed and danced through the long Atlantic swell, to slide up the stern ramp like an orange cataract in reverse. 'Dick!' A stentorian shout from Gus MacDonald, doyen of the Marine Laboratory's fishing-gear men and as mighty a hunter as the boatswain himself, shook me out of my reverie. He was in the act of cutting a post-smolt salmon out of the upper cod end. It was the first of two in only our second haul at the northern end of the Faeroe-Shetland Channel, and this success was followed the day after by hauls of ninety-nine post-smolts and then fifty-five more in the eddying surface waters at the south-east end of the Wyville-Thomson Ridge.

Among the wild fish we could make out some, a little above average size, with the fin deformities conferred by artificial rearing. Like the wild fish, they also had been feeding opportunistically on the small crustaceans, mainly amphipods and euphausiids, and post-larval fishes that we knew from our plankton sampling to be the most nourishing mouthfuls in the area. However, their stomachs did not look as full as those of their wild shoalmates. Our suspicions were confirmed when, back at the laboratory, we were able to tell, from the narrow spacing of the growth rings on their scales, that they were still having trouble learning the skilled trade of mid-water visual predator. They were growing at not much more than a third of the rate of the wild fish, their above-average size a legacy of their earlier life of plenty rather than of their current well-being.

It looked as though a lot of the additional mortality suffered by hatchery-reared fish occurred in the first weeks after they left the river because, over that critical period, they were spending too long being small and vulnerable. We knew from the work of our Norwegian colleague Lars-Petter Hansen, who had sampled hatchery released salmon later in their lives at sea, that the survivors eventually achieve the marine growth-rates of wild salmon. Some of the hatchery fish were carrying microtags, tiny slivers of magnetized stainless steel inserted into their nose cartilages at the time of their release. We were able to read their batch numbers and to get in touch with the scientists who had tagged the young fish. Two of the fish had been tagged in fresh water as parr, one in northern Spain and the other in Wales. Interesting as these records were, however, they told us nothing about how long the fish had been at sea.

All the remaining fish had been tagged as smolts at head of tide in Ireland and had been at sea for some six to eight weeks. By calculating the distances between their points of release in Irish estuaries and their sites of capture in the Faeroe-Shetland Channel, we were able to estimate their rate of migration. To our surprise, it was slower than the speed of the Shelf Edge Current in which we caught them. Either they had 'hung about' off the coast of Ireland after their release or, more likely, from what we had seen of the hydrography from satellite photographs, they were feeding in the eddy systems that swirl along the edge of the continental shelf. Another revelation from the Irish tag returns, and we recovered twenty-six of them in two successive years, was that the smolts that had left together from individual rivers like the Shannon and the Bundorragha tended to stay together at sea. We had suspected that something like this might happen from scattered observations in the past, and that sequences of tags applied at the smolt stage would be recovered around the same time in adults returning to home waters, but this was the first time that we had direct evidence that the home-water origins of salmon persisted at sea. It was a worrying observation because it highlighted the vulnerability of particular populations of salmon to directed or accidental interception by trawlers fishing at the surface.

Cruise by cruise, we pieced together the early lives of salmon at sea. On one of them we were joined by the Norwegian biologist Dr Jens-Christian Holst, who had long been concerned that post-smolt salmon which had passed down long fjords heavily utilized for the cage-rearing of salmon, would acquire dangerously high infestations of sea lice. In early June, fifteen miles off the coast of southern Norway, we caught the first example of such an unfortunate fish. This specimen was carrying over a dozen pre-adult lice, a level of infestation which Dr Holst's subsequent experiments and research cruises would show to be almost certainly lethal by mid-July.

Later, off the north-east coast of Scotland, we were able to follow the smolt migration from leaving brackish water on the Cromarty Firth to entering full-strength sea water in the inner and outer Moray Firths. In the Cromarty Firth, wind-blown insects were as important as marine organisms in the diets of the young fish but, once out of the Moray Firth, they fed on the usual range of crustaceans and post-larval sand eels, *Ammodytes marinus* Raitt. Unlike their adult parents, young sand eels spend long periods of time in dense shoals close to the surface, and it is here that they fall prey to post-smolt salmon. The results of our sampling suggested that the small shoals of post-smolts operated on the edges of the concentrated masses of sand eels. For the young salmon, maintaining contact with fellow members of their own shoal, who could well have been kith and kin, was an important part of their defence against predation, and it would be all too easy for that contact to be lost when foraging in the porridge of a sand-eel throng.

We were still, however, a long way from having all the pieces of the jigsaw puzzle. All of the wild post-smolts we had caught were growing rapidly through their period of greatest vulnerability thanks to the abundance of their food supplies in the waters around Scotland. It looked as though the changes in marine climate that had reduced the survival rates of salmon at sea were exerting their malign influence later in the lives of the growing fish. It is a problem that has yet to be solved directly. However, in recent years salmon biologists on both sides of the Atlantic have noticed unusual checks in the scales of some returning adult salmon, in the part of the scale corresponding to the end of the first summer at sea. Given that these fish were survivors, it could well be that greater interruptions in growth were suffered by those who were subjected to more severe food shortages in the patchy world of the open ocean. Failing to 'grow to avoid mortality' can, for a salmon, so easily be a ticket to the warm inside of a marine mammal's stomach.

We had greater success in estimating the risks of inadvertent interception by fisheries for other species. For some years, salmon anglers had worried that the large Danish industrial fishery for sand eels in the North Sea posed a danger to post-smolt salmon. Fortunately for the young salmon, the fishery is for adult sand eels and takes place on the seabed in areas where there is a good match between the grade of the sand and the biological requirements of the sand eels. Our observations were that post-smolt, and probably also older salmon, spend most of the long hours of summer daylight close to the surface. Their chances of encountering a sand-eel trawl were therefore minimal – a conclusion borne out by the observation that the very limited by-catch in the sand-eel fishery consists either of demersal fish, which spend the whole of their time close to the seabed, or mid-water fish, which spend part of their time there. A more justifiable concern was that the scale of the sand-eel fishery was so enormous that it posed a threat to the food supply of the young salmon. That it could not do so directly was confirmed by our observation that the post-smolt salmon feed, not on the adult sand-eels against which the fishery is directed, but on their post-larval progeny. So long as the numbers of the young sand eels are limited not by the numbers of eggs laid by their parents but by the user-friendliness of their early environment, all is well. How long such a reassuring state of affairs will continue, only time will tell.

We knew, from the work of our Norwegian colleagues, that the relatively new fashion of fishing for mackerel at the surface posed a proven risk to post-smolt salmon. We also knew that a large mackerel fishery had developed in the international zone of the Norwegian Sea. Known from its shape as the Banana Hole, it is an area outside the Exclusive Economic Zones (EEZs) of the European Union and of those coastal states which have sensibly remained independent of the EU. It extends in a great arc from halfway between Iceland and Norway into the high latitudes of the Arctic Ocean. Regulation of fishing in the Banana Hole is in the hands of the North-East Atlantic Fisheries Commission (NEAFC), an organization of which all of the relevant coastal fishery authorities are members. In August 2000, the fishery protection duties in the area were discharged by the ships of the Scottish Fisheries Protection Agency (SFPA), a remarkable private navy commanded by a distinguished post-captain poached from the Senior Service and run to the highest standards by master mariners of the Merchant Marine.

Friendly rivalry between the deck officers of the Merchant Navy and the seaman officers of the Grey Funnel Line has existed since the separation of the two services that followed the Dutch wars of the seventeenth century. The holder of a master's ticket tends to be be a professional seaman first and foremost, and the 'pusser routine' (spit-and-polish) of the Royal Navy have limited attraction for him. In the same way, the informal appearance of all too many members of the modern Merchant Service can appal even the most easy-going product of Britannia Royal Naval College. A first recorded rapprochement was to come at the hands of Captain Maurice Suckling RN, who arranged for his nephew, Horatio Nelson, to spend a year learning the hard trade of seamanship aboard an East Indiaman. So impressed was the young Nelson by the morale and efficiency of his new shipmates that he was most reluctant to rejoin his parent service. The value of basing discipline on mutual respect for the skills of officers and ratings, rather than the threat of brutal punishment, was to remain with him until his death at Trafalgar.

By a curious coincidence, Nelson's youthful experience of leadership by example was to find an early expression through his service with the Fishery Protection Squadron of the Royal Navy, still the proud guardians of the great man's razor. The Scottish Fisheries Protection Agency guards no such relic but, in the smartness of its officers, who have the privilege of wearing the same insignia of rank as their Royal Naval equivalents, and the efficiency of its ships, it embodies a tradition which would have gladdened the heart of Captain Maurice Suckling.

The oldest of the four-ship SFPA flotilla but, until this year, still its pride, FPV *Westra* was also the best sea boat. She was an early member of the 'Island' class, and the shape of her hull was based on that of the frigates and corvettes that had coped best in northern waters during the grim days of the Battle of the Atlantic. Battleship grey with the merest hint of blue, she wore her twenty-five years lightly and, as we nosed out of Leith's Victoria Dock to enter the broad, navigationally tricky waters of the Firth of Forth, no one was prouder of his command than her young captain, Alastair Beveridge. To port lay the former Royal Yacht, *Britannia*, the remains of her glamour marred by her dressing lines of bunting that sagged as if supporting so much washing. Ahead lay a long steam to sixty-five degrees North where, at the southern end of the Banana Hole, over thirty Eastern Bloc factory trawlers, mainly of the venerable 'Atlantik' and 'Super-Atlantik' classes, were fishing intensively for blue whiting, *Micromesistius poutassou* (Risso), and mackerel.

Run-down rusty hulks, their catches trans-shipped back to Russia and their supplies replenished at sea, these trawlers and their friendly but dull-eyed crews were shortly to become our familiar companions. By chance, the first factory trawler to loom out of the swirling fog banks was not one of the veterans but *Murman II*, a gleaming vision with white upper works and royal blue hull, the very colours of poor *Britannia*. A wet and bumpy ride across in the rigid-hulled inflatable and we were welcomed aboard to hear her story. Her friendly skipper, bear-like in sweatshirt and sandals, presided over a spotlessly carpeted wheel-house bristling with the most up-to-date fish-finding and navigational equipment.

Built in Chile, engined by Caterpillar, owned by a Norwegian company and crewed by Russians, *Murman II* was a floating United Nations flying the flag of the British Virgin Islands. Blue whiting had been her quarry that day. Fished for at around 200 metres in tows that could last anything from six to twelve hours, the blue whiting were being stored prior to processing in great tanks of refrigerated sea water. The crew had not seen any salmon since July, when they had been fishing at the surface for mackerel in the same area. Then they had sometimes seen them passing along the processing line. Most were mackerel-sized post-smolts which no doubt ended up as frozen fillets along with the rest of the catch. Every so often a bigger one would appear, however, and was set aside to provide some variety in the crew's diet.

To our surprise, the less fortunate crews of the Russian-owned trawlers were equally forthcoming and, by the time we had to leave the area, we had an excellent qualitative picture of the occurrence of salmon at the southern end of the Banana Hole. It was clear that both post-smolts migrating north through the intensive surface mackerel fishery and returning adults on their way south were highly vulnerable to inadvertent capture. Once through the danger zone, they were relatively safe. Our results were based on the casual reports of Russian crew members spotting fish tumbling along processing lines. The value of these was the information they provided on timing and sizes. Just how many were caught we could not know at the time but, the following year, the extrapolated results from Norwegian research vessels, fishing in June and July and with the same gear as the factory trawlers, suggested that several million might be wasted in this way. Given that fish from particular catchments seemed to stick together, the effects on the future fortunes of even the greatest of Atlantic salmon rivers could no longer be guaranteed.

This was hardly reassuring news, yet such was the state of affairs when I retired as director of the Freshwater Fisheries Laboratory and ended, temporarily at least, my obsession with the lives of fish and fishermen – an obsession which had endured, undiminished, for over forty years. None the less, despite the many man-made hazards that currently face Atlantic salmon and the irrational modern attitude to wildlife management which has allowed predators like grey seals and cormorants to proliferate, the story of these ancient fishes is very far from over. Most of their new problems are marine ones which threaten their well-being by reducing the numbers and sizes of the fish which return to spawn. Given the resolve, all but the global changes in marine climate are solvable. Indeed, some measurable progress has been made through the controls which have been placed on the more indiscriminate of the high-seas and coastal fisheries and through the efforts which are being made to identify and reduce losses caused inadvertently by fisheries for other sorts of fish.

A less tractable problem is presented by a salmon-farming industry growing ever larger in its increasingly desperate pursuit of economies of scale. Drug treatments are now available which can reduce sea-louse infestations among caged salmon to the point where they are much less of a threat to the wild salmon and sea trout whose habitat they pollute. Much less progress has been made in preventing escapes and thereby reducing the threat to the genetic integrity of wild fish populations. Whether predatory seals, birds and fish attracted to fish from cages are a real danger to wild salmon and sea trout is not known, and neither do we know whether the strong smell of caged salmon, fish food and faeces interferes with the olfactory cues which guide wild salmon and sea trout to their natal rivers. What we fear, but have yet to experience, is the introduction or *in situ* generation of a disease or parasite problem which could spread to wild fish, much as the freshwater parasite *Gyrodactylus salaris* has devastated salmon stocks in parts of Norway. *G. salaris* was introduced accidentally with Baltic smolts resistant to it, and so serious were its effects that the Norwegian authorities were reduced to poisoning the fish in the affected rivers in attempts to eliminate the intruder.

The better news is that the places which are most important of all to salmon and sea trout – the river systems in which they breed and rear their young – are, if anything, in better order now than they have been for some years. Especially is this true of estuaries where, as a result of the decline in heavy industry and improvements in sewage treatment, catchments like those of the Clyde and the Tyne are now supporting self-sustaining salmon populations. Even my beloved Chess is now regularly stocked with salmon parr and a few return to the lower Thames every year, stopped only by lock gates from returning to their point of stocking.

Looking back to the earlier history of the salmon's encounters with modern Man, the only instances where salmon stocks became extinct were when industrial and domestic pollution and the construction of dams and weirs denied the fish access to their spawning and nursery grounds. That this lesson has been learned, and limited remedial action taken, is good news indeed. Our challenge now is to match and surpass it with comparable stewardship of the seas on which the wild salmon depend for their food.

WELLS REVISITED

It had been a good run ashore even if, in Tromsø, the beer was the equivalent of £5 a pint and the dried fish which accompanied it had the consistency of dry newspaper. Early the following morning, the Royal Norwegian Research Ship *Johan Hjort* stole away from the quay, swung in the tideway and, gathering speed, began her long passage north through the island-girt leads to her lonely way point at seventy-four degrees North where, at the Polar Front, the North-East Atlantic meets the Arctic Ocean, and whales and salmon sport in the banquet that is the Arctic summer.

No longer a civil servant at the beck and call of politicians and their spin doctors, I had joined my Norwegian colleagues Jens-Christian Holst and Marianne Holm in my new capacity as Research Director of the Atlantic Salmon Trust (AST), an international charity devoted to the well-being of Atlantic salmon and sea trout. My role aboard the *Johan Hjort* was to lend my recent experience of Atlantic salmon at sea and the moral support of the Trust to

The Royal Norwegian Research Ship Johan Hjort

the expedition. We were keen to add to our knowledge of the northward movement of post-smolt salmon from the British Isles and Norway to their summer feeding grounds in the gyres that swirl to the north and west of a seabed feature known as the Vøering Plateau. We were also keen to catch pre-adult salmon so that we could fit them with tags that stored data on the temperature and depths through which they had swum. During our cruise, which would take us far above the Arctic Circle before finally ending at the *Johan Hjort*'s home port of Bergen, we would enjoy the company of hundreds of salmon and the sight of killer, minke and sperm whales. Like us, they were making the most of the constant daylight. If for us the novel blaze of the midnight sun was a sight to lift the spirits, for the fish and whales it was the source of energy on which their livelihood depended.

Back in Tromsø lay the sunken, and still unsalvaged, wreck of the once mighty German battleship *Tirpitz* and soon, far to our east off the North Cape, we would remember the men of the elegant *Scharnhorst*, prefixed 'lucky' by her ship's company but finally to be overwhelmed by Admiral Sir Bruce Fraser's Home Fleet in 1943, just four short years before my brother and I first saw the sea at Wells.

'Have we reached England yet?' asked Peter as our host Henry Aldridge's boat kissed the muddy bank of Sluice Creek and we scrambled excitedly ashore amidst the samphire and sea lavender. The British Anzani outboard that had taken us to this terra nova had only a thousandth of the power of the mighty in-line diesel that drove the 1,800 tons of the *Johan Hjort* through the long swells of the Greenland Sea. Over 1,200 miles and fifty-six years now separated the grizzled figure on the bridge from the wide-eyed, well-muddied savage at Wells-next-the-Sea, but within days of my return from the *Johan Hjort*, it would be to Wells that I would take Freda and our grown-up sons, John and Neil, and their friends.

There have been many changes since that first visit when the quay still played host to railway trucks and before the great North Sea surge of 1953 smashed through the sea bank and devastated the Corsican pines of the Holkham Estate. The clinker-built whelk boats, every one of which did its bit at Dunkirk, have gone, but crabs, shrimps and line-caught mackerel are landed and, every so often, large Dutch sailing barges thread their way up the channel to maintain Wells' proud reputation as a working port.

One of the few advantages of growing older is that one does not need as much sleep. Long before the family were up, I would walk down to the harbour to live again those distant days with few visitors and even fewer motor cars. Sometimes my only companions were the harbour master, Bob Smith, on his rounds and Myrtle French feeding the ducks and swans with lacy fragments of batter from her family's renowned fish and chip shop. Both were children of Wells and both, like me, enjoyed reminiscing about the great days when the little town and its dunes and salt marshes played host to the gentlemen gunners.

Much has changed since 1947, as the buildings of Wells attest. Only the new lifeboat station fulfils its original function, although the old one on the quay still serves with honour as the harbour office. The granary, with its covered gantry, has been turned into expensive apartments and the fishermen's own public house, the Shipwright's Arms, is now a private house. As to the old whelk-houses at the head of the harbour, most are still in service but the greatest and most historically interesting, the mecca of our childhood visits to Wells, has long since been demolished to make way for a yacht park.

Enough, however, remains for Wells still to be Wells. As we walked the harbour to collect our samphire the way we always had, the redshank still shrieked their alarms and little terns still plopped fearlessly in pursuit of young sand eels. As we rinsed the soft mud from the samphire's white roots, Meadow Brown, Gate Keeper and Painted Lady butterflies skittered among the sea lavender. Despite floods, wars, the loss of its railway (it now has two splendid miniature ones of its own) and summer invasions on a scale unheard of half a century ago, Wells remains true to itself.

The author (for once without the hated glasses) and Peter (right)
at Wells-next-the-Sea in the late 1940s

What is remarkable, about not just Wells but the whole of the north Norfolk coast from Brancaster to Salthouse, is not that there have been changes – such are inevitable when the restless North Sea is the final arbiter – but that the essential character of this long-shore paradise has not been lost. Flint still faces even the newest cottages and many of the old have been sensitively restored and now look good for another 200 years. We had rented a wonderfully restored, pantiled cottage with a walled garden scented by rose-mary and lavender. While the boys and their friends explored the salt marshes, played boules at the Buttlands and learned to like warm English bitter, Freda and I shopped for locally caught sea trout at Burnham Market and visited again the sea bank where my father's first goose fell to his Coster ten-bore. How thrilled he would have been to know that the grey geese have returned to Wells in numbers not seen since the nineteenth century. Now a new generation of my family has taken north Norfolk to its heart and counts a car journey of hundreds of miles a small price to pay for spiritual renewal.

THE BROKEN GATE

Take away the tweeds and the bespoke shotguns and there is not that much difference between a smart pheasant shoot in Devon and chimpanzees catching colobus monkeys in the forests of central Africa. Once the high-status individuals (the guns) are in position, the lower-status ones (the beaters) move forward to flush the quarry to their superiors. As the biologists Desmond Morris and Richard Dawkins have observed, it is not just that Man's ancestors were apes – he still is an ape with the built-in potential to take pleasure, like his closest living relative the chimpanzee (with which he shares over 98 per cent of his DNA), in hunting animals and gathering fruits and other handy stores of energy.

Not many modern men are invited to smart pheasant shoots and few women spend much of their waking hours gathering blackberries and hazelnuts. For most, work and shopping are the inadequate substitutes, and both vary a great deal in their capacity to satisfy the instincts that make such activities attractive. I was one of the lucky ones whose play in childhood and work in later life followed, in its broad outlines, the path of the hunter gatherer. As the years passed, so the steam locomotives and small fishes pursued during the years of childhood and adolescence became the wild geese, lobsters and salmon of adulthood. The eye that grew to recognize the ways in which mechanical engineers developed one locomotive design from another was the same eye that was later trained to see the evolutionary relationships between animals, and the energetic ones between the key biological processes of growth,

The longshoreman with his Tolley eight-bore and two greylags

survival and reproduction. Truly could it be said in my case that the child was the father of the man.

If Man's drives and instincts are those of a great ape, his capacity to indulge them to excess is the result of his powers as a toolmaker and his ability to communicate his thoughts to others through oral and written language. Through the latter he has been able not just to exchange ideas with his contemporaries, but to benefit from the experience of previous generations in a way that is not possible for other animals. The process is called cultural evolution and it moves at a much faster pace than its biological counterpart.

Biologically speaking, the men who bequeathed the 30,000-year-old masterpieces of Lascaux were identical to the Europeans of today. As predators and inhabitants of a lost Eden, however, they were very different. The only main specializations, other than the discovery of fire-making, which distinguished them from their fellow killers were the ability to use weapons to strike at a distance and to co-operate when hunting prey too large to be overcome by an individual. Baboons can do the former, wolves and hunting dogs the latter. Although in every respect modern men, our immediate ancestors were at that time as fully integrated into the biological communities around them as any other large mammal. Above all, their capacity to grow and reproduce was directly linked to the rate at which they were able to gather and store food.

Times of plenty were times of high levels of infant survival – there would even be enough to spare for those enfeebled by age and injury. When food was short, more babies and children died and nothing was left for the halt and the lame. It was a ruthless discipline for men with the same propensity to care for others as ourselves, but it guaranteed the food supply of succeeding generations. By subsequently learning to grow crops and to corral his prey, Man was able to move the point at which food became limiting to a higher threshold. However, once the new threshold had been reached, the old rules governing survival to adulthood still applied and the instincts built into his hunter-gatherer brain still served him well, despite the wastefulness of early slash-and-burn agriculture.

So far as European fishery resources were concerned, there was no real equivalent of that first agricultural revolution. Admittedly, monks had first reared carp in freshwater ponds in the medieval heyday of Catholic Christendom and the extensive culture of shellfish, especially of oysters, also has a long history. However, most North European fisheries are based on the exploitation of wild stocks for human consumption or to provide protein sources for pigs, chickens and salmon in intensive culture. Wild fisheries are sustainable so long as catches do not exceed what fishery scientists call 'the exploitable surplus'. If too many fish or shellfish are caught too early in their lives, the future of the fishery is compromised. As we have seen, the economic effect of such over-exploitation may be blunted for a time by increases in the value of individual fish. Ultimately, however, it is the fishermen, and those whose shore jobs depend upon them, who suffer when catch-rates are no longer sufficient to meet the costs of fishing. As with predator–prey relationships generally, it is the abundance of the prey that controls that of the predator.

There is, however, one great difference between modern fishing and the artisanal subsistence activity from which it developed. It is that the instinctive fine tuning which prevented the latter from straying beyond the exploitable surplus is not naturally built into the former. Without external regulation, the economic viability and biological basis of modern fisheries, including intensive aquaculture, are doomed to fluctuate between feast and famine. Furthermore, the lesson of recent history is that periods of famine tend to be longer than those of plenty.

When a previously abundant species such as cod is seriously depleted, other species with shorter generation times, like squid and fish of low commercial value, may proliferate and delay the recovery of the desired quarry. Even when the collapsed stock eventually recovers, processing facilities that have long since gone out of business may take many years to re-establish themselves. Most of these lessons had been learned by the middle of the last century and any who had doubts about the benefits which would accrue from controlling the fishing power of modern fleets had them removed by the manifest improvements to fish stocks which followed the two world wars.

It was after the second of these wars, and a few years before Peter and I caught our first lamprey, that Michael Graham, a veteran of the First World War who was then working in operational research at Bomber Command Headquarters, put the finishing touches to what was to become the greatest book written on the philosophy of commercial fisheries. It was called *The Fish Gate*, and if its title seems strange to those brought up in our present secular age, it was a reference to one of the gates of Jerusalem by the Prophet Zephaniah. By declaring that 'the Fish Gate is open', Graham summarized his book in one line. He looked to a post-war world in which large accumulated stocks of fish would be managed according to the theory of fishing he and his scientific and fishermen colleagues had begun to work out but in ways which were neither biologically wasteful nor detrimental to the instinctive desire to secure his quarry that is hard-wired into the brain of Man the hunter.

It was a wonderful dream and, in his yearning for a better post-war world based upon rational planning, Graham was far from alone. For a time, his dream was realized. The years of respite granted to the stocks by the war had benefited home- and distant-water resources alike. The fishermen who had served in Harry Tate's Navy and survived came home to opportunities not seen since the 1920s. The scientists who worked alongside them used the tools of operational research to develop new approaches to understanding how the effects of fishing – and changes in the capacity of the environment to support fish production – combined to affect the size and composition of both current catches and future spawning stocks.

Sadly, neither new-found scientific understanding, nor the resolution required in the light of its first warnings, were sufficient to confer stability on the fishing industry. Feast followed famine with dreary predictability as the stocks, first of pelagic species like herring and then demersal ones like cod and haddock, oscillated wildly and to the great detriment of the prosperity of coastal communities. No longer could the sons of the Harry Tate's men look forward to a secure future.

Now there is hardly a fishery resource in the North-East Atlantic which is not either fully exploited or considerably over-exploited – and this at a time when the large Spanish fleet of modern fishing vessels is about to gain access to the North Sea grounds which are the mainstay of what is left of the British fishing fleet. Given this latest humiliating consequence of the loss of fishery sovereignty suffered under the terms of our entry to what was once called the European Common Market and is now the EU, it is understandable but regrettable that successive British administrations have been so fiercely protective of the intensive rearing of salmon in coastal waters and rainbow trout in inland ones. Only now is the realization dawning that an over-sized fish-farming industry – an industry which is dependent upon over-exploited species of fish to feed its progeny and which dumps its copious wastes in restricted waters ill-adapted to receive them – makes neither biological nor economic sense. It is, in fact, little more than ecological asset-stripping.

Such was the bleak news which greeted me at the end of a career which had included service in five fishery laboratories and in countless research and fishing vessels. In more pessimistic moments, I suppose I might perhaps have been tempted to dismiss the whole exercise as the mere indulgence of the childhood interests of one in whom the instinct of the longshore hunter remained unusually strong. But that temptation never arose.

That, in fact, it had all been so much more than a mere indulgence was largely thanks to the instrinsic 'crackability' of most of the fishery problems with which I was faced. This was especially true of the fisheries for lobsters, crabs, salmon and sea trout. All are fisheries in which – either because of the informal arrangements between fishermen to divide the seabed between them (coastal fisheries for crabs and lobsters) or direct ownership of the fishing rights (domestic fisheries for salmon and sea trout) – fishermen have a direct interest in both regulating themselves and in excluding others. None of these 'stakeholder' fisheries is free from problems beyond the control of its practitioners but, equally, none has been driven to the point of collapse by directed fishing alone.

It has been suggested that only by making the fishermen of today directly responsible for the well-being of the resources they exploit, can the stability of their industry be regained. For fishery resources exploited over large areas of sea by trawl and seine nets, achieving such an outcome will always be difficult. However, the example of Iceland, which was the first European nation to extend its fishing limits and then to match the extension with the rational management of its fleet, shows what can be done when fishery sovereignty is intact and when the list of species to be managed is relatively short. The task is much more difficult where the exclusive fishing zone is not the preserve of a single nation and there are many species with diverse needs to conserve. It is under these great handicaps that the EU Common Fishery Policy operates.

The CFP has had a few successes, notably the regulations that permitted the partial recovery of herring stocks following their post-war collapse. On the whole, however, its story has been one of irresolution and compromise and of promoting wider access to resources already over-exploited by both fishermen and grey seals. Perhaps its worst feature is its use of single-species quotas to control the rate at which fish die. All too often the effect has been the opposite of that intended, as large quantities of fish whose quota has been reached are discarded overboard in pursuit of those whose quota is still open. Almost all of the unfortunate discards die either immediately or shortly after their return to the sea. How much this cruel and wasteful practice may have contributed easily available food to grey seals, and thereby helped to fuel the enormous growth in their numbers, no one yet knows.

Half a century's experience of fish and wildfowl is a very short time in the context of the ten millennia since the end of the last ice age. It is nevertheless quite long enough to be certain that Man's capacity to deplete wild populations, and to damage the environments upon which they depend, is at least as great as the worst of natural catastrophes. If that is the bad news, the good is that the powers of recovery which have evolved to cope with such events are also the key to making Man's harvests, and the disposal of his wastes, infinitely sustainable. It has been my good fortune mainly to have been associated with activities in which the capacity to forgive, which is built into the natural world, has not been grossly abused. Most rewarding of all has been to live long enough to see the grey geese return to my beloved Wells in numbers undreamed of by my father's generation. Even more pleasing is that the lesson that disturbance and over-exploitation was the reason for their long absence has been so well learned that the great skeins are now secure for all time to bring joy to naturalist and 'fowler alike.

And what of the sublime purpose for which Man was created: the pursuit of wildfowl on the foreshore with the great guns of the late nineteenth century? In the outhouse, a brace and a half of pinkfeet cool in the frost of late winter; above the Raeburn, the thirty-four-inch barrels of an eight-bore made in 1878 dry slowly, along the top rib the proud inscription, 'J. & W. Tolley, St Mary's Square, Birmingham'.

> Home is the sailor, home from sea,
> and the hunter home from the hill.

Robert Louis Stevenson, 'Requiem' (1887)

'Those in peril'

SELECT READING LIST

(All titles published in London unless otherwise stated)

ABERDEENSHIRE

Adair, Elizabeth, *North East Folk. From Cottage to Castle*, Paul Harris
Publishing, Edinburgh, 1982

Allan, John R., *Green Heritage*, Ardo Publishing, Buchan, 1991

Porter, William A., *Tarves Lang Syne*, Maxiprint, York, 1996

Whiteley, Archie W. M. (ed.), *The Book of Bennachie*, The Bailies of
Bennachie, Aberdeenshire, 1976

Whiteley, Archie W. M. (ed.), *Bennachie Again*, The Bailies of Bennachie,
Aberdeenshire, 1983

FISH AND FISHERIES

Burgess, G. H. O., *The Curious World of Frank Buckland*, John Baker, 1967

Davidson, Alan, *Mediterranean Seafood*, Penguin, 1972

Davidson, Alan, *North Atlantic Seafood*, Macmillan, 1979

Festing, Sally, *Fishermen, A Community Living from the Sea*, David &
Charles, Newton Abbot, 1999

Frost, W. E. & Brown, M. E., *The Trout*, Collins, 1967

Graham, Michael, *The Fish Gate*, Faber & Faber, 1953

Hardy, A. C., *The Open Sea*, Part 1, *The World of Plankton*, Collins, New
Naturalist Series, 1956

Hardy, A. C., *The Open Sea*, Part 2, *Fish and Fisheries*, Collins, New
Naturalist Series, 1959

Jones, J. W., *The Salmon*, Collins, 1959

Kurlansky, Mark, *Cod*, Jonathan Cape, 1998

Lawrie, Alistair, Matthews, Helen, and Ritchie, Douglas (eds.), *Glimmer of Cold Brine. A Scottish sea anthology*, Aberdeen University Press, Aberdeen, 1988

Lee, A. J., *The Directorate of Fisheries Research. Its Origins and Development*, MAFF, 1992

Maitland, P. S. and Campbell, R. N., *Freshwater Fishes*, Harper Collins, 1992

Mills, Derek (ed.), *Salmon at the Edge*, Blackwell Science, Oxford, 2003

Mitford, William, *Lovely She Goes!*, Pan, 1969

Schweid, Richard, *Consider the Eel*, University of North Carolina Press, Chapel Hill, 2002

Summers, David W., *Fishing off the Knuckle. The Fishing Villages of Buchan*, Centre for Scottish Studies, Aberdeen, 1988

Wheeler, Alwyne, *The Fishes of the British Isles & North-West Europe*, Macmillan, 1969

Wigan, Michael, *The Last of the Hunter Gatherers. Fisheries Crisis at Sea*, Swan Hill, Shrewsbury, 1998

Youngson, Alan and Hay, David, *The Lives of Salmon*, Swan Hill, Shrewsbury, 1996

GENERAL NATURAL HISTORY

Attenborough, David, *Life on Earth*, Collins, 1979

Redfern, Ron, *Origins. The Evolution of Continents, Oceans and Life*, Cassell, 2000

NELSON AND THE NAVY

Bennett, Geoffrey, *Naval Battles of the First World War*, Pan, 1968

Coleman, Terry, *Nelson*, Bloomsbury, 2001

Jolly, Rick, *Jackspeak the Pusser's Rum*, Palamanando Publishing, Cornwall, 1989

Mallalieu, J. P., *Very Ordinary Seaman*, Granada, 1956

Southey, Robert, *The Life of Nelson*, 1813

Trotter, Wilfrid Pym, *The Royal Navy in Old Photographs*, J. M. Dent, 1975

RAILWAYS

Casserley, H. C., *The Later Years of Metropolitan Steam*, D. Bradford Barton, Cornwall, [n.d.]

Foxell, Clive, *Chesham Shuttle*, Clive Foxell, Chesham, 1996

Ransome-Wallis, P. (ed.), *The Concise Encyclopaedia of World Railway Locomotives*, Hutchinson, 1959

Simpson, Bill, *The Aylesbury Railway*, Oxford Publishing Co., Hersham, 1989

ST ANDREWS

Forrest, Catherine, *Living in St Andrews*, St Andrews University Library, St Andrews, 1996

Lamont-Brown, Raymond, *St Andrews*, Sutton Publishing, Stroud, 1996

THE SEAS

Byatt, Andrew, Fothergill, Alastair, and Homes, Martha, *The Blue Planet*, BBC, 2001

Kunzig, Robert, *The Restless Sea*, W. W. Norton, New York, 1999

WELLS, WILDFOWLING & GUNS

BB, *The Sportsman's Bedside Book*, Eyre and Spottiswoode, 1948

Bishop, Billy, *Cley Marsh and its Birds*, The Boydell Press, Woodbridge, 1983

Colquhoun, John, *The Moor and the Loch*, William Blackwood & Sons, 1888

Cringle, Mike, *The Gamekeeper's Boy*, The Larks Press, Norfolk, 2001

Crudgington, I. M. and Baker, D. J., *The British Shotgun. Vol. I 1850–1870*, Barrie & Jenkins, 1979

Crudgington, I. M. and Baker, D. J., *The British Shotgun. Vol. II 1871–1890*, Ashford Buchan & Enright, Leatherhead, 1992

Hastings, Max, *Outside Days*, Pan, 1989

Hawker, Lt. Col. P., *Instructions to Young Sportsmen in all that relates to Guns & Shooting*, Herbert Jenkins, 1922

Hesketh Prichard, H., *Sport in Wildest Britain*, William Heinemann, 1921

Marchington, John, *The History of Wildfowling*, A & C Black, 1980

Martin, Brian P., *Wildfowl of the British Isles and North-West Europe*, David & Charles, Newton Abbot, 1993

Millais, J. G., *The Wildfowler in Scotland*, Longmans, Gregg and Co., 1901

Parker, Eric (ed.), *Colonel Hawker's Shooting Diaries*, The Derrydale Press, Mississippi, 1990

Pooley, Graham, *The Port of Wells. 1100 Years of History*, Graham Pooley, Norfolk, 1992

Savory, Alan, *Norfolk Fowler*, Geoffrey Bles, 1953

Storey, Neil R., *The North Norfolk Coast*, Sutton Publishing, Stroud, 2001

Wentworth Day, J., *Sporting Adventure*, Harrap & Co., 1937

PICTURE CREDITS

The author and publisher would like to thank the following for their kind permission to reproduce images in *The Longshoreman*:

CEFAS, Lowestoft for the photograph of the Tellina and crew (page 169) © Crown copyright; Captain Paul du Vivier RN, Chief Executive, Scottish Fisheries Protection Agency for the final image in the book (page 330) © Crown copyright; Dundee Archive services for the portrait of D'Arcy Thompson (page 246); David Hay for the photograph of a leaping salmon (page 279); Professor Ian A. Johnston of the Gatty Marine Laboratory for the portrait of McIntosh (page 107) and McIntosh's drawings of the shanny (page 120); Angus MacKinnon for his illustration of HMS *Birmingham* (page 298); Mirrorpix for Torrey Canyon (page 131) and Rockall Island (page 269); St Austell China Clay Museum for the painting of Carclaze Pit by Elliot (page 166).

All other pictures are from the author's own collection.